A
CONTEMPLATION
UPON FLOWERS

In the cottage of the rudest peasant;
In ancestral homes, whose crumbling towers,
Speaking of the Past unto the Present
Tell us of the ancient Games of Flowers.

In all places, then, and in all seasons,
Flowers expand their light and soul-like wings,
Teaching us, by most persuasive reasons,
How akin they are to human things.

And with childlike, credulous affection,
We behold their tender buds expand—
Emblems of our own greater resurrection,
Emblems of the bright and better land.

Henry Wadsworth Longfellow
"Flowers," 1839

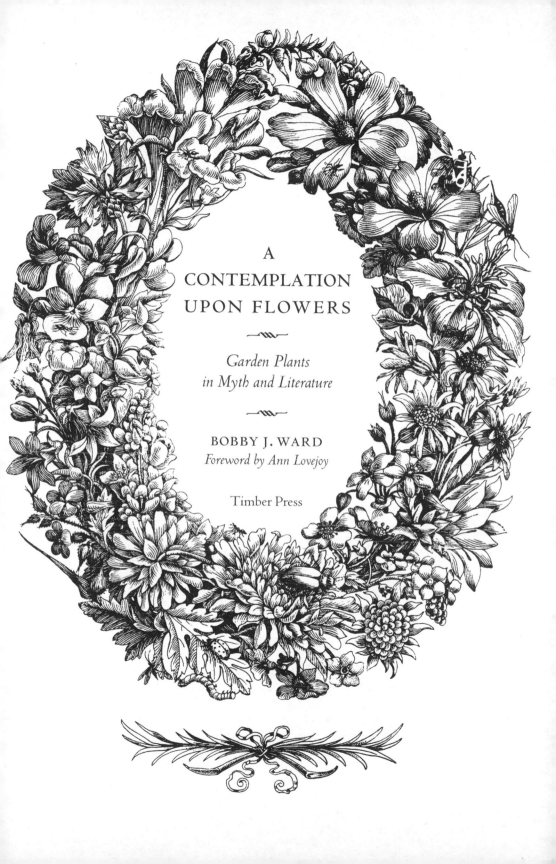

A
CONTEMPLATION
UPON FLOWERS

*Garden Plants
in Myth and Literature*

BOBBY J. WARD
Foreword by Ann Lovejoy

Timber Press

Paperback edition published in 2005 by
Timber Press, Inc.
The Haseltine Building
133 S.W. Second Avenue, Suite 450
Portland, Oregon 97204, U.S.A.

www.timberpress.com

Designed by Susan Applegate

Printed in Hong Kong

Library of Congress Cataloging-in-Publication Data

Ward, Bobby J.
    A contemplation upon flowers : garden plants in myth
and literature / Bobby J. Ward.
        p.   cm.
    Includes bibliographical references and index.
    ISBN 0-88192-727-9
    1. Flowers in literature. 2. Flowers—nomenclature
(Popular) 3. Flowers—Folklore. I. Title.
PN56.F55W37   1999
809'.93364—DC21   98-51438
                                                              CIP

CONTENTS

FOREWORD

by Ann Lovejoy

OME gardeners are content to let the flow of seasons dictate the hours that may be spent in the company of plants. Those of us who are entirely smitten with our favorite pastime, however, are loath to leave the world of flowers behind. When autumn winds howl, when winter winds whistle, when spring rains fall, we spend our enforced leisure among the ever-blooming, paper gardens found in books.

Bobby Ward is most certainly this sort of gardener. His book provides yet another delicious escape from work, weather, and worry for gardeners who love to trace the history of plants, following their travels and remarking their shifting popularity. Unless the luxury of a well-stocked library is close at hand, however, such researches can present a significant challenge. Here gathered in a single, fruitful volume is a harvest many years in the gleaning. Curious, questing, and determined, Ward has trailed, traced, and tracked down hundreds of literary references to many of our favorite flowers. His patience provides us all with the rich reward of hours of sumptuous browsing in this storehouse of reverie.

Like many dedicated gardeners (who tend to be closet if not overt sensualists), I am an inveterate bathtub reader. As the foam-stained pages of my copy of this bulky manuscript can attest, *A Contemplation Upon Flowers* has been a frequent companion. I can assure you that this is a perfect choice to read while soaking in scented suds.

It is also splendid for reading aloud during the unscenic parts of a lengthy road trip, when this volume's conjured images most pleasantly replace repetitive miles of strip malls. How infinitely more attractive it is to follow the threads of myth and legend, poem and song as they weave a floral tapestry whose origins stretch back for centuries.

The only problem with this approach is that it may necessitate frequent bookstore stops. Very often, an enticing reference directs us forcibly back to the original. Although Ward has done so much research for us, his snippets often rather urgently demand our return to the source. Reading about Tess gathering lords and ladies from a shady bank whilst chatting with Angel Clare made me spend almost an hour searching out my battered old copy of *Tess of the D'Urbervilles*, which I then reread with renewed pleasure and a new eye for its floral references.

The chapter on asphodels sent me to the book shelf again, this time for *Sonnets from the Portuguese* and *Paradise Lost*. Reading about buttercups led to an impromptu Gilbert and Sullivan song session. The study on hardy cyclamen revealed that these exotic-looking flowers attracted the notice of John Ruskin, whose naturalist writings inspired the Pre-Raphaelite revolution. And Ruskin's name led to the particular treat of finding a new (if old) book that will be treasured, for I had never before read his *Proserpina*, a discourse on wild flowers in which it is easy to detect influences that affected Victorian art and music as much as literature.

Though *A Contemplation Upon Flowers* will probably appeal most strongly to imaginative visionaries, practical gardeners will find plenty to think about as well. The history of well-traveled flowers, such as the dahlia, tells us that they can be grown in a surprising variety of places. Those who are not themselves versed in Latin will be warned that *Myosotis palustris* prefers damp sites, and trusting beginners fooled by finding forget-me-nots sold as annuals will discover that their perpetual plants are not miracles, but reliable perennials.

The daring may be led to try to brew heather beer, as was done in

old Scotland, while the craft-minded may wish to make handsome little brooms with their callunas or carve burl pipes from the root knots
of their arboreal ericas. The determinedly uncrafty (like myself) will
be content simply to know that such matters might be attempted. But
I would not recommend that anyone, no matter how crafty, attempt
to assuage a scorpion's bite with a handful of fern bracken.

Perhaps my favorite part of this delightful book is the chapter entitled "Saints of the Spade." Lots of folks are familiar with St. Fiacre,
the Irish monk whose blobby little statues decorate many an herb garden. Who, however, knows as much about St. Phocas of Sinope, or
St. Adelard of Normandy, or St. Urban of Langres, a protector against
frosts and blights? These are clearly patrons toward whom we might
fruitfully direct certain petitionary prayers.

A very thorough bibliography allows the reader to retrace Ward's
researches. This final section makes it comparatively simple to follow
up on the book's myriad divinations, employment that guarantees
countless hours of playful exploration in the gardens of the past.

ACKNOWLEDGMENTS

SEVERAL people offered special assistance in selecting information for this book. I owe appreciation to Edna Whitson of Cockfosters, England; Gretl Meier of Santa Ynez, California; the staff of the Royal Horticultural Society Lindley Library at Vincent Square in London; the staff of the New York Botanical Garden Library in the Bronx; and Bill Burt and Jeff Beam of the Botany Library of the University of North Carolina at Chapel Hill. Fellow gardeners Nancy Goodwin, Judy Glattstein, and Bobby Wilder steered me to some literary resources.

Two people, now deceased, I want to acknowledge for the advice they gave along the way of this book's creation. Indefatigable gardener William Lanier Hunt, an active nonagenarian at the time, was frequently available for lunch and long conversation and allowed me the use of books from his vast personal library. And Linda Mitchell Lamm offered continuous warm encouragement.

A third great influence on this project has also passed away. Over the last two decades, all horticultural activities in North Carolina sooner or later intersected with J. C. Raulston of the North Carolina State University Arboretum. This book is no exception. J. C. initially suggested a collaboration between Timber Press and me and went so far as to offer his house for an appointment with Neal Maillet, acquisitions editor at Timber Press. Neal's unabashed enthusiasm encouraged this project to its fruition.

Roy Dicks was the initial reader and snipper of thorns from each draft chapter. He also served as a personal reference librarian, double-checking dates and text accuracy, both on-line and in the stacks. Barbara Scott, Larry Thomas, Tom Stuart, and Steve Whitesell provided helpful suggestions that framed the early stages of the manuscript. Mike Chelednik read all drafts, scanned for nomenclatural problems, and assisted in illustration selections. The last draft of the manuscript was critically read by Brian Bixley, Nancy Goodwin, John Grimshaw, Ron Maner, and Al Simpson. The assistance of all these individuals is greatly acknowledged.

Ellen Kussow of Timber Press provided a keen eye and competent editing skills, which measurably improved and shaped the final manuscript.

I have made every effort to quote only from the public domain. Some excerpts, particularly translations, are not quoted from the work of the original author. In these cases I have tried to give accurate credit and source information. I apologize for any errors in attribution.

INTRODUCTION

I sing of brooks and blossoms, birds, and bowers,
Of April, May, of June and July flowers.
I sing of Maypoles, hock carts, wassails, wakes,
Of bridegrooms, brides, and of their bridal cakes.
I write of youth, of love, and have access
By these to sing of cleanly wantonness.
I sing of dews, of rains, and, piece by piece,
Of balm, of oil, of spice, and ambergris.
I sing of times trans-shifting, and I write
How roses first came red and lilies white.
I write of groves, of twilights, and I sing
The court of Mab and of the fairy king.
I write of hell; I sing (and ever shall)
Of heaven, and hope to have it all.

Robert Herrick
"The Argument of His Book," *Hesperides*, 1648

FLOWERS and gardens have been objects of contemplation for writers from the beginning of recorded history. The beauties and mysteries of the horticultural world inspired early historians, clerics, and poets everywhere, from Greece (Homer in the ninth century BCE), Rome (Virgil in the first century BCE), India (Bhasa in the third century CE), China (Wang Wei in the eighth century), Germany (Walafrid Strabo in the ninth century), and England (Chaucer in the fourteenth century). By the end of the sixteenth century, printing technology had allowed the dissemination of literature in ways that were previously impossible; and for the next three centuries, poets, essayists, dramatists, and novelists continually incorporated references to plants into their works.

Before the mid-1800s, writers and scholars had to be very well read to know where to search for botanical references in literature, as no compilations were generally available. Very early concordances to the Bible existed in manuscript form, but the first mass-printed edition was not published until 1737. The first concordance to Shakespeare did not appear until 1790. Although not intended for garden and plant citations alone, Samuel Johnson's *Dictionary* of 1755 and John Bartlett's *Familiar Quotations* of 1855 provided major steps toward a method for finding horticultural literary references.

The last quarter of the nineteenth century saw a sudden surge in the publication of collections of literary references to plants and flowers. In many cases, excerpts were combined with plant legends, lore, and the language of flowers, a Victorian system that assigned specific meanings to flowers. One name that appeared frequently in those collections was Rev. Hilderic Friend (an authority on earthworms), who in 1884 wrote the extensive, albeit rambling, two-volume *Flowers and Flower Lore*. A more organized yet undiscriminating work, *Plant Lore, Legends, and Lyrics* by Richard Folkard, Jr., was published the same year and has served as the basis for latter-day students of plant lore to search out obscure legends, myths, and traditions. Some of the books published within this genre in the 1800s were references designed to interest young women in the scientific principles of botany. These books often included illustrative excerpts of poetry. A popular title was Margaret Plues's *Rambles in Search of Wild Flowers*, the third edition of which was published in 1879.

In the twentieth century and especially in its last few decades, a number of researchers have gathered literary quotations about both gardens in general and individual flowers. However, only a few anthologies have been published exclusively on plants and flowers in literature. Of the latest examples, *Wild Flowers in Literature*, published in London in 1934, is a detailed and exacting collection by Englishman

Vernon Rendall, who spent a lifetime collecting the material. Regret-tably his work and most others like his are out of print and general-ly unavailable in standard reference libraries. With grateful acknowl-edgment of the help received from these aforementioned works, I have compiled this book to fill the gap for a current anthology.

I encountered several difficulties in collecting citations for this book. First, many of the common plant names used in classic litera-ture have not survived into current usage, or they refer to plants whose taxonomy and botanical nomenclature have changed. For example, Shakespeare at times invented names and sentiments for plants to suit his purposes; thus, the plant names he used often are not used in mod-ern English. To Shakespeare, the buttercup (*Ranunculus acris*) was known as the cuckoo-bud, and lady's smock (*Cardamine pratensis*) was the mead-ow flower. The same phenomenon applies to plant names used by other writers of centuries past. Wordsworth knew the lesser celandine (*Ranunculus ficaria*) or buttercup. Charles Dickens wrote of gowan (*Bel-lis*) or daisy. Tennyson mentioned the cuckoo-pint (*Arum maculatum*), now the arum or lords and ladies. And Ben Jonson knew the paunce (*Viola*), now the pansy.

Regrettably, the old folk names are rarely used in written form any-more. But Roy Vickery of the Plant Lore Society and the Natural History Museum in London has cataloged the folk names currently spoken in different regions of the British Isles and published his find-ings in *A Dictionary of Plant Lore* (1997). It was a valuable reference for me because it identified vernacular names for plants and flowers that might otherwise have gone unidentified. You will find many of those names—many with rich histories—in this volume.

A second difficulty I encountered was verifying that a recognizable common name was applied to the same plant from writer to writer. For example, to the botanist, the western European common daisy is *Bellis perennis*. However, the Marguerite daisy refers to the same *Bellis* spe-cies. Thus, when Wordsworth in "To the Daisy" wrote, "Bright flower! whose home is everywhere"; or when John Clare pronounced, "The

daisy is a happy flower"; or earlier still when Chaucer wrote how
Queen Alceste "was turned into a Daisie," I have generally chosen not
to quibble about the genus or the species to which the authors were
referring. I simply accepted at face value the common English or its
translated name and relied on the context to help readers identify the
plant botanically and horticulturally, if necessary. Try to tease apart
the following lines from "An Ode" from *Il Pastor Fido* by Sir Richard
Fanshawe and attempt to append Latin binomials to them—you will
see why I chose to relegate much of the nomenclatural maze to the
Index of Plants:

> The commonwealth of flowers in its pride
> Behold you shall;
> The lily queen, the royal rose,
> The gilliflower, prince of the blood!
> The courtier tulip, gay in clothes
> The regal bud;
> The violet, purple senator,
> How they do mock the pomp of state,
> And all that at the surly door
> Of great ones wait.

Another factor to consider regarding plant nomenclature in literature is that today's plant names do not necessarily represent the same
plants they did hundreds of years ago. Thus "lilies of the field" (Matt.
6:28) in the Bible is generally thought to mean any pretty flower,
though *Sternbergia lutea* and *Anemone coronaria*, neither of them true lilies,
are likely possibilities. Lilies have been mentioned by many writers:
Chaucer wrote of a "lilye whyte," Milton of "Ladon's lilied bank,"
and Virgil of "handfuls of lilies to scatter." Because the name lily was
formerly more generally and widely used in English than it is now, it
refers to scores of plants: daylily, lily of the valley, Madonna lily, waterlily, atamasco lily, arum lily, and canna lily, just to name a few. Per-

haps we should simply accept the following quatrain from "Maud" (1854) by Alfred, Lord Tennyson:

> What is it? A learned man
> Could give it a clumsy name,
> Let him name it who can,
> The beauty would be the same.

An example of name changes or perhaps simple misidentification occurs in René Rapin's *De Hortorum Cultura* (1665, translated by James Gardiner in 1706) in which the following curious lines appear:

> Late from Japan's remotest Regions sent,
> Narcissus came array'd in purple Paint,
> And numerous Spots of yellow stain the Flower,
> As richly sprinkled with a golden Shower:
> The radiant tinctures may with tap'stry vye,
> And proudly emulate the Tyrian dye.

During the seventeenth century the name narcissus was used for any amaryllid. Rapin appears to be describing *Nerine sarniensis*, a member of the Amaryllidaceae from coastal South Africa, not Japan as had been widely assumed in Rapin's time.

Attempting to sort out these nomenclatural discrepancies for the modern horticulturist, I have kept within easy reach of my computer *The Dictionary of Plant Names* (Coombes 1994), *Hortus Third* (1976), and similar books. However, even with these resources, we cannot really know what species Sir Walter Scott had in mind when he wrote, "And purple throatwort, with its azure bell." I accept it as a campanula and guess that it might have been *Campanula rotundifolia* based on its distribution, Scott's description of it, his location, and the probable location of this particular subject.

The selections I found to build this collection confirm the four categories that Richard Folkard described as recurring themes in literary associations with flowers. One is the mythological theme, such as the story of the narcissus springing from the dead body of the beautiful youth Narcissus. The historical significance of flowers and plants is a second theme, for example, the white and red roses that stood for the two rival houses of York and Lancaster, respectively. Third, ecclesiastical and religious symbolism has for centuries been incorporated in literature through horticultural connotations. And the fourth theme is the use of flowers in poetic associations. This last category I further divide into the more specific themes of fairy tales and superstitions, love and relationships, observations of nature, and the language of flowers.

The language of flowers has an intriguing history in itself. Lady Mary Wortley Montagu, wife of the British ambassador in Constantinople, is credited with initiating the language of flowers in western Europe. On 16 March 1718 she sent a letter to a friend in Britain: "I have got for you as you desire, a Turkish Love-letter, which I have put in a box, and ordered the Captain of the Smyrniote to deliver it to you with this letter." Lady Mary explained the meaning of the pieces she included: roses, jonquils, cloves, straw, and miscellaneous non-botanical items. This collection of flowers in the form of a tussie-mussie reached its peak of popularity in the Victorian era. By the early 1800s the French court had "le langage des fleurs" and the Germans were using a "Blumen-Sprache." The sentiments expressed by a particular flower could vary from language to language; in Turkish and Arabic the sentiment assigned to a flower originated not from the flower itself but from the ability of its name to rhyme with other words. As Lady Mary said, "There is no flower without a verse belonging to it; and it is possible to quarrel, reproach, or send letters of passion, friendship, or civility, or even of news without even inking your fingers."

The mixture of literary citations in this book comes from authors as diverse as the plants they wrote about. And tradition is often the author of the bits of legend and plant lore that I have thrown into the collection. This anthology is not intended to be exhaustive, nor is it a systematic coverage of the great works of literature. In choosing to include a work, I was guided by how the flower was described or mentioned or if it struck a particular chord with me, not by the fame or infamy of the author. Thus you will find in these pages moralists, agnostics, Buddhists, Christians, and an array of others. Poets, scholars, doctors, farmers, preachers, and gardeners are represented here. Some, like Chaucer and Shakespeare, have been known to me since my high school days; others, like John Clare and Leticia Elizabeth Landon, were uncovered only by leafing through faded pages in secondhand bookstores and libraries in the United States and England. Some writers I found to have been keen observers of flowers and nature. Others appeared to have had little firsthand experience, but their literary skill brought out the significance of the plants.

This collection contains familiar pieces and obscure ones. Gathering these selections was much like tending a small garden when one has a vast amount of seed and plants with which to fill the limited space. What to sow and what to toss? Here may be missing the familiar "Rose is a rose is a rose," but included is a touching piece by Irishman Tadhg Dall O'Huiginn of the late sixteenth century: "Her cheeks outblushed the rose." Unfortunately, this collection contains only a limited sampling from works outside the Western tradition. A great deal of my work for this book involved deciphering colorful plant names from older texts, and I would have been completely uncertain of the terrain had I tried to interpret references in non-Western sources. Since the process of compiling the final manuscript necessarily involved great powers of selectivity, I maintained a focus in support of the large part of this book that has to do with the mythological significance of flower names that have come to the English language (and the international language of botanists) from Latin and Greek.

For the sake of brevity, I was forced to weed out references to vegetables, herbs, and trees, limiting the collection to ornamental garden plants and their wild flower allies. The most difficult decision to make was to toss on the mulch pile a plethora of botanically vague references, wonderful and desirable as they may be. For example, Oscar Wilde wrote lovingly in "Athanasia" of a flower produced by a seed found clutched in the hand of the mummified body of a young girl uncovered in an Egyptian pyramid and brought to England:

> But when they had unloosed the linen band
> Which swathed the Egyptian's body,—lo! was found
> Closed in the wasted hollow of her hand
> A little seed, which sown in English ground
> Did wondrous snow of starry blossoms bear,
> And spread rich odours through our springtide air.

I do not know the identity of the plant that contained the "starry blossoms," but his description suggests several possibilities.

For most of the citations I relied on traditional and conventional bibliographic resources. However, one resource unavailable until the 1990s was both a time-saver and revelation: the Internet. Through the on-line searching of numerous literary databases it is possible, for example, to go to a site on the World Wide Web and read the *Chanson de Roland* in both medieval French and English. Electronic forms of the complete works of Shakespeare and the Bible in multiple versions and translations are posted on the Internet. Best of all, key-word searching capabilities programmed into the web sites greatly facilitated my research. Paramount among these new electronic opportunities is Project Gutenberg, a vast undertaking with the goal of making available on the Internet 10,000 complete books by the year 2001. I used the project's CD-ROM, which was produced in 1994 at the early, though already well-developed, stages of construction. This mammoth enterprise helped me considerably to reduce trips to libraries and efficiently gather research material from web sites.

I am somewhat embarrassed to admit that during the course of my research the book at times became incidental to my ferreting. I would often dart excitedly from one plant reference to another like a puppy out of a box, and at other times I would spend an entire evening diverted down the path of some unrelated topic like a dog with a new bone to chew on. It might be days before I returned to my initial focus. I was buoyed along by the excitement of finding an uncommon work of literature or a perspective previously unknown to me, something by Walter Landor, Robert Herrick, or John Ruskin, or from a Korean folktale, a Greek myth, or a Chinese poem.

Unwittingly, I gained an education through my work. I never knew that Chaucer read and wrote French and Latin as well as "Chaucerian" English, that he lived in France and Italy, and that he wrote numerous poems besides *The Canterbury Tales*. I learned of saints associated with flowers and gardening and found that violets were the symbol of Napoleon's confederate conspirators. I discovered the "Eros of Garden" by Oscar Wilde and found in Giacomo Meyerbeer's opera *L'Africaine* that the tropical American plant, the manchineel (*Hippomane mancinella*), incongruously grows on Madagascar. I found that in Aztec mythology the dahlia symbolizes the Serpent Woman and that the spirit of Buddhist followers rests in the lotus, which takes them to nirvana. I learned that many of the generic names for plants were traditionally taken from goddesses and nymphs, a practice initiated by Linnaeus after Virgil's *Georgics* in which Phyllodoce—one of the fifty sea-nymph daughters of Nereus and Doris known as the nereids—and her companions "carded fleeces of miletus dyed a rich glass-green colour." I discovered Victorian women poets, and I read Shakespeare with a new set of eyes and ears. Try the following account of Proserpina's flowers from act 4 of *The Winter's Tale* (1610):

> O Proserpina,
> For the flowers now, that frighted thou let'st fall
> From Dis's waggon: daffodils
> That come before the swallow dares, and take
> The winds of March with beauty; violets dim,
> But sweeter than the lids of Juno's eyes
> Or Cytherea's breath; pale primroses,

> That die unmarried, ere they can behold
> Bright Phoebus in his strength, a malady
> Most incident to maids; bold oxlips and
> The crown-imperial; lilies of all kinds,
> The flower-de-luce being one. O! these I lack
> To make you garlands of, and my sweet friend,
> To strew him o'er and o'er.

I understood the pain so eloquently expressed by Milton at the loss of his friend and wept at his description of the floral bier in "Lycidas (A lament for a friend drowned in his passage from Chester on the Irish Seas, 1637)":

> Ye valleys low, where the mild whispers use
> Of shades, and wanton winds, and gushing brooks,
> On whose fresh lap the swart Star sparely looks,
> Throw hither all your quaint enamell'd eyes,
> That on the green turf suck the honied show'rs,
> And purple all the ground with vernal flow'rs;
> Bring the rathe Primrose that forsaken dies,
> The tufted Crow-toe, and pale Gessamine,
> The white Pink, and the Pansy freakt with jeat,
> The glowing Violet,
> The Musk Rose, and the well-attir'd Woodbine;
> With Cowslips wan, that hang the pensive head,
> And every flower that sad embroidery wears;
> Bid Amaranthus all his beauty shed,
> And Daffadillies fill their cups with tears,
> To strew the laureat hearse where Lycid lies.

The education I gained during my searches frequently resulted from a casual reference or a serendipitous find, such as the name Rev. Francis Kilvert, which led me to his diaries and the folk of his parish in England near the Welsh border. A citation about hellebores from

Elizabeth Lawrence led me to Erasmus Darwin, the grandfather of Charles, then to his book-length poem *The Botanic Garden*, and months later in London to finding that the Wellcome Institute for the History of Medicine had just published a concordance to it. And, most appropriately for this book, I learned the intriguing fact that the word anthology itself has a floral basis. It comes from the Greek for "flower gathering" and was used as the name for small collections of choice poems, or "flowers of verse."

A Contemplation Upon Flowers is meant to be an equally delightful education for readers. Pick it up and read it front to back, or dart about from flower to flower. Gathering this anthology has been genuine fun and a joy, giving me many hours of personal pleasure. I hope readers will find pleasure among these pages as well.

BOBBY J. WARD
Raleigh, North Carolina

Brave flowers—that I could gallant it like you,
And be as little vain!
You come abroad, and make a harmless show,
And to your beds of earth again.
You are not proud; you know your birth:
For your embroider'd garments are from earth.

You do obey your months and times, but I
Would have it ever Spring:
My fate would know no Winter, never die,
Nor think of such a thing.
O that I could my bed but view
And smile, and look as cheerfully as you!

Henry King, Bishop of Chichester
"A Contemplation Upon Flowers," 1657

ACANTHUS

Climbing sprays of lithe acanthus

The thirty species of the perennial *Acanthus* were originally distributed across the Mediterranean, into Asia Minor, and throughout tropical Africa and Australia. Acanthus flowers are typically purple and
white, and the plant may have leaves up to 1 meter (3 feet) long. Among
the most widely grown species are *Acanthus mollis*, the common acanthus; the sharply spiny *A. spinosus*; and from the Balkans, *A. balcanicus*
(*A. hungaricus*).

The genus name comes from the Greek *akanthos*,
referring to the plant's thorny leaves and bracts.
The shape of the leaves is also the inspiration for
the common name bear's breeches, which is derived from *brank-ursine*, a Middle Latin word
meaning bear's claw. *Acanthus mollis* of Europe
and northern Africa is the bear's breeches
mentioned in literature.

Because of the sharp spines on some
species of acanthus, the English
herbalist John Parkinson cataloged it with thistles (*Carduus*)
in *Paradisi in Sole, Paradisus Terrestris*
(1629). About the "garden beares
breech," he wrote: "The leaves
of this kinde of smooth thistle (as it is accounted) are almost as large as the leaves of
the artichoke, but not so sharp
pointed, very deeply cut in and

gashed on both edges, of a sad green and shining colour on the up-perside, and of a green underneath, with a great thicke rib in the mid-dle." Regarding the "wilde or prickly beares breech," he noted, "This prickly thistle hath divers long greenish leaves lying on the ground, much narrower [than] the former, but cut in on both sides, thicke set with many white prickes and thorns on the edges."

In the language of flowers—a code widely used in Victorian Eng-land that assigned a meaning to each of hundreds of flowers—acan-thus signified the fine arts and artistic skill. Greek and Roman archi-tects used the leaf of acanthus, probably *Acanthus spinosus* or *A. mollis*, as a model for the designs on the capitals of Corinthian columns. Cal-limachos, an Athenian sculptor and architect of the fifth century BCE, was the first to create the acanthus design. Stories vary as to how the design came about, but most commonly it is told that the young daugh-ter of Callimachos died while her father was designing pillars in Cor-inth. In her memory he set a basket of flowers on her grave, using a tile to keep the wind from blowing them away. The next time he visi-ted the grave, an acanthus plant had sprung up around the basket, and where its leaves reached the top of the tile, they bent back around its corners. Callimachos was so struck by the beauty of the combination of the tile's architecture and the plant's foliage that he introduced the design of the curled acanthus leaves as a capital for his columns. In some versions of the story, a nurse placed the tile in the graveyard in memory of another child. Both the Greeks and Romans liberally used the representation of acanthus on buildings, furniture, and clothing.

With the advent of its architectural use, the plant began appearing in literature. Virgil wrote of an acanthus design embroidered on Helen of Troy's mantle. In "The Cup" from *Idyll*, translated by Sir William Marris, the Greek poet Theocritus described a scene carved on a cup. While two vixens romp and three lovers talk, a young boy weaves a cricket cage

And never minds his wallet or his vines,

So happy is he in his basket-work.
And all around the cup are climbing sprays
Of lithe acanthus.

In the seventeenth century John Milton in book 4 of *Paradise Lost* (1667) included acanthus among the plants in the bower of the Garden of Eden:

[the] roof
Of thickest covert was inwoven shade,
Laurel and myrtle, and what higher grew,
Of firm and fragrant leaf; on either side
Acanthus, and each odorous bushy shrub,
Fenced up the verdant wall.

Acanthus appeared in at least two poems in the nineteenth century. Alfred, Lord Tennyson, described an acanthus wreath in part 7 of *The Lotos-Eaters* (1832):

To hear the dewy echoes calling
From cave to cave thro' the thick-twined vine—
To watch the emerald-color'd water falling
Thro' many a woven acanthus-wreath divine!

His poem recalls Homeric legend, in which people who ate from the lotus tree would forget family, friends, and the desire to return home; rather, they would live in ease and luxury, but idleness, in lotus land. And the Victorian poet Jane Francesca Wilde, mother of Oscar Wilde, used acanthus in "The Poet's Destiny." Writing in the late 1800s under the pen name Speranza, she considered the solitude of the poet in a world enriched by but oblivious to him. The poet wears a crown of thorny acanthus as a symbol of the responsibility he bears:

All that is noble by his word is crown'd,
But on his brow th' Acanthus wreath is bound.

Eternal temples rise beneath his hand,
While his own griefs are written in the sand;
He plants the blooming gardens, trails the vine—
But others wear the flowers, drink the wine.

For its early role as a monument to a little child, for its presence in
the most idyllic gardens, for its common character, rugged and tough,
acanthus has remained in our minds, hearts, and gardens for some
twenty-five centuries.

ACONITE

Earth has borne a little son

The winter aconite with its bright yellow, solitary flower is *Eranthis hyemalis*, a member of the buttercup family, Ranunculaceae. Although native to southern Europe, *E. hyemalis* has become naturalized in some areas of Britain and North America. An herbaceous plant with underground tubers, *Eranthis* consists of only seven species, but they are widely recognized in the Northern Hemisphere as a sign of the returning spring. Winter aconite grows well under deciduous trees with well-drained soil and is suitable for rock gardens. One aconite of garden origin, *E. hyemalis* Tubergenii Group, contains the late winter-blooming 'Guinea Gold', which has exceptionally large flowers that are deep yellow and persistent. The group of cultivars is named after the Dutch bulb nurseryman, C. G. van Tubergen.

The name *Eranthis* derives from the Greek words *ar* and *anthos* and means spring flower. The species name *hyemalis* comes from Latin and means, paradoxically, of winter. The winter aconite was once charmingly called New Year's gifts in Lincolnshire and yellow star in other regions of England. In the 1800s it was also known as Roman soldiers or Romans in the belief that it only grew in England where Roman soldiers had shed their blood. How the common name winter aconite began to be applied is uncertain. The name is shared with its lethal botanical relative *Aconitum*, the monk's-hood or wolfbane. The names aconite and *Aconitum* are derived from the Greek word *akoniton*, meaning without dust or without struggle, suggesting that at one time *E. hyemalis* was also used (or misused) as a pharmaceutical or poison. Authors have freely interchanged the name aconitum for winter aconite and other buttercup family members, thereby casting a shadow on this bright, winsome herald of spring. For example, Henry Lyte's translation of Rembert Dodoens's *Niewe Herball or Historie of Plantes* (1578) calls it "little yellow woolfes-bane."

Exactly which aconite various authors intend is sometimes not clear. Shakespeare, who knew plants and their attributes, mentions an aconitum and its deadly powers in act 4 of *Henry IV, Part II* (1597). King Henry charges his son Thomas to counsel his brother Hal, heir to the throne, to accept the burden of kingly responsibility to prevent the family bond from breaking and their claim to the crown from disintegrating:

> Learn this, Thomas,
> And thou shalt prove a shelter to thy friends,
> A hoop of gold to bind thy brothers in,
> That the united vessel of their blood,
> Mingled with venom of suggestion—
> As, force perforce, the age will pour it in—
> Shall never leak, though it do work as strong
> As aconitum or rash gunpowder.

Another Englishman, John Clare, knew woodland plants and described winter aconite fairly accurately: "With buttercup-like flowers that shut at night,/Its green leaf furling round its cup of gold." John Gerard described winter aconite as "the small yellow winter-wolfsbane, whose leaves come forth of the ground in the dead time of winter, many times bearing snow on the heades of its leaves and flowers; and the deeper the snow is, the fairer and larger is the flower, and the warmer the weather is, the lesser is the flower." Ben Jonson, perhaps referring to its dangerous relative, said in his play *Sejanus his Fall* (1603), "I have heard that aconite being timely taken, hath a healing might/ Against the scorpion's stroke."

In Greek mythology aconites (or more likely *Aconitum*) are associated with Hecate, a Titan and the queen of hell. As the goddess of the dead in the underworld, she taught witchcraft and knew the names and uses of all herbs. Aconite also plays a part in some versions of the story of Medea. Taking revenge on her husband, Jason, she poisoned her children with a brew from the aconite. In *Metamorphoses* (translated by John Dryden et al.) Ovid told of Medea's use of aconite and went on to give the dark origin of the plant, in which Heracles, or Hercules or Alcides in Roman mythology, left the nether regions to carry out his twelfth labor and took with him Cerberus, the watchdog of hell. The three-headed animal furiously spat venomous saliva that fell to the earth and from which sprouted the aconite.

> Medea, to dispatch a dang'rous heir,
> (She knew him) did a pois'nous draught prepare,
> Drawn from a drug, long while reserved in store,
> For desp'rate uses, from the Scythian shore,
> That from the Echydnaean monster's jaws
> Derived its origin, and this the cause.
>
> Through a dark cave a craggy passage lies
> To ours ascending from the nether skies,
> Through which, by strength of hand, Alcides drew
> Chained Cerberus, who lagged and restive grew,
> With his bleared eyes our brighter day to view.
> Thrice he repeated his enormous yell,
> With which he scares the ghosts, and startles hell;
> At last outrageous (though compelled to yield),
> He sheds his foam in fury on the field;
> Which, with its own and rankness of the ground,
> Produced a weed by sorcerers renowned,
> The strongest constitution to confound—
> Called Aconite, because it can unlock
> All bars, and force its passage through a rock.

Notwithstanding the occasional literary misassociation of the winter aconite with aconitum, poisonous spittle, and the underworld, it is a beloved plant for those who can get it to grow and a welcomed bright spot in the garden after the winter storms. A. M. Graham in "The Aconite" considered the flower the special tender offspring of Mother Earth. He is clearly describing eranthis:

> Earth has borne a little son,
> He is a very little one,
> He wears a bib all frilled with green
> Around his neck to keep him clean.
> Though before another Spring
> A thousand children Earth may bring
> Forth to bud and blossoming—
> Lily daughters, cool and slender,
> Roses, passionate and tender,
> Tulip sons as brave as swords,
> Hollyhocks, like laughing lords,
> Yet she'll never love them quite
> As much as she loves Aconite:
> Aconite, the first of all,
> Who is so very, very small,
> Who is so golden-haired and good,
> And wears a bib, as babies should.

In my own garden in North Carolina that consists of heavy clays formed during the Triassic period, I have had less than desirable results growing and maintaining winter aconites. Perhaps I need the faith that Louise Beebe Wilder expressed in *Adventures with Hardy Bulbs* in which she calls them the "most venturesome" of plants "that spring from most unpromising looking small dark irregular-shaped tuberous roots that often, when received, resemble bits of old dried wood."

ADONIS, ANEMONE
& PULSATILLA
My sweet love's flower

Adonis, anemone, and pulsatilla are botanically and horticulturally distinct, but they are related as members of the same botanical family, Ranunculaceae, the buttercup family. Because they are somewhat similar visually, writers have used their names interchangeably. Their taxonomy and nomenclature have changed over the years, but the three plants do have distinctive characteristics and histories.

The genus *Adonis* includes more than twenty species; the two most widely grown are *A. vernalis* with yellow petals and *A. annua* with crimson petals. The common name pheasant's eye refers specifically to *A. annua*, but it is also applied to the genus as a whole. The amur adonis, *A. amurensis*, which blooms in late winter with long-lasting flowers, is most often found in the garden in its various cultivated forms, typically with yellow or orange petals that are bronze on the outside. Adonis is distributed in Eurasian temperate regions.

The genus name evokes the Greek youth Adonis, who has come to symbolize youthful male beauty. According to Greek mythology, Aphrodite (Venus in Roman mythology), the goddess of love and beauty, was in love with Adonis. But Ares was in love with Aphrodite and jealously plotted the "accidental" death of Adonis. Disguised as a wild boar, Ares killed Adonis, but Aphrodite transformed the wasted body of her lover into a flower. The blooming of the flower and its fading away have become a metaphor for the premature death of Adonis. The myth of the flower's life resulting from the youth's death is also symbolic of the course of a plant's life.

In some versions of the story, an anemone sprang from the tears of Aphrodite as she wept over the bleeding body of her lover. In other versions, the flower sprang from the blood of Adonis. In *Venus and Adonis* (1593), Shakespeare described a purple flower, perhaps referring to Adonis's blood rather than to the crimson color of the actual adonis flower. Perhaps he borrowed the image from earlier writers, because his description suggests *Fritillaria meleagris*:

> By this boy that by her side lay kill'd
> Was melted like a vapour from her sight,
> And in his blood, that on the ground lay spill'd
> A purple flower sprung up, chequer'd with white,
> Resembling well his pale cheeks and the blood
> Which in round drops upon their whiteness stood.
>
> She bows her head, the new-sprung flower to smell,
> Comparing it to her Adonis' breath;
> And says, within her bosom it shall dwell,
> Since he himself is reft from her by death;
> She crops the stalk, and in the breach appears
> Green-dropping sap, which she compares to tears.
>
> . . . .
>
> "Here was thy father's bed, here in my breast
> Thou art the next of blood, and 'tis thy right;
> Lo, in this hollow cradle take thy rest;
> My throbbing heart shall rock thee day and night;
> There shall not be one minute in an hour
> Wherein I will not kiss my sweet love's flower."

—ᴍ—

The genus *Anemone* includes about 120 species of perennial herbaceous plants distributed worldwide in temperate regions, particularly in the Northern Hemisphere. The early spring-blooming wood anemone of Europe is *A. nemorosa*, from which numerous cultivars have been selected. *Anemone coronaria* of southern Europe has solitary flowers of blue, pink, scarlet, or white. *Anemone ×hybrida*, which probably includes in its parentage *A. hupehensis* from central China, provides flowers in late summer to autumn in a range of colors, from magenta through pink to white. North American species include *A. narcissiflora*, which grows near the timberline in the Rocky Mountains and is also a Eurasian species, and *A. parviflora*, which is found from the Alaskan tundra to Colorado. *Anemone tuberosa* grows in the chaparral deserts of the southwestern United States.

Traditionally, the genus name *Anemone* was derived from the Greek word for wind, *anemos*, leading to the common appellation windflower, which has been passed along for centuries. However, the word anemone may derive from the Semitic *na'aman*, meaning "the handsome one." This theory may help explain the confusion that has existed between the plant names anemone and adonis. Various tellings of classical myths give the adonis and the anemone the same origin, but another myth is singular to anemone and explains the common name windflower. Anemone was a nymph favored by Zephyr, the god of the gentle west wind. This favoritism angered Flora, the goddess of flowers and spring, so that she banished Anemone from the court and transformed the young nymph into an early-spring-blooming flower. Zephyr abandoned the flower to the brusque north wind, Boreas, whose rough caresses in the form of chilly spring winds caused the anemone to bloom and fade quickly.

The Greek poet Bion, writing ca. 100 BCE, subscribed to the myth of Aphrodite's (or Venus's) tears as the origin of the anemone—also introducing the theory of the rose originating from Adonis's blood—in "Death of Adonis," translated in 1759 by John Langhorne:

> Thus Venus griev'd—the Cupids round deplore;
> And mourn her beauty, and her love no more.
> Now flowing tears in silent grief complain,
> Mix with the purple streams, and flood the plain.
> Yet not in vain those sacred drops shall flow,
> The purple streams in blushing roses glow:
> And catching life from ev'ry falling tear,
> Their azured heads anemonies shall rear.

René Rapin in *De Hortorum Cultura* (1665, translated by James Gardiner in 1706) told a similar story of Venus and Adonis. The four parts of Rapin's grand Latin poem consist of a parade of garden plants and a litany of myths and legends associated with them:

> But which of all the cruel Deities,
> Exposed the Garden's pride, Anemonies.
> Beauties so tender, to such rigid dooms,
> For storms to shake and snow to hide their blooms.
> Fame does another diff'rent Fable tell,
> That when Adonis, Venus's Darling, fell;
> Slain by the savage Boar this Flow'rs relief,
> Charm'd the fair Goddess, and suppress'd her Grief.
> For while what's mortal from his Blood she freed,
> And Show'rs of Tears on the pale Body shed;

> Lovely anemonies in order rose
> And formed a purple Pall, and veil'd her sacred Woes.

In "The Triad" of 1828 William Wordsworth described the dance muse wearing a windflower, a flower that also symbolizes the muse (the nine muses were inspirational goddesses of the arts and sciences, the daughters of Zeus and Mnemosyne, the goddess of memory):

> But her humility is well content
> With one floweret (call it not forlorn)
> Flower of the Winds, beneath her bosom worn
> Yet more for love than ornament.

Thomas Edward Brown, writing in the late 1800s in part 7 of "Cleve-don Verses," also described the anemone as a humble, shy flower:

> In Norton Wood the sun was bright,
> In Norton Wood the air was light,
> And meek anemonies,
> Kissed by the April breeze,
> Were trembling left and right.

John Clare knew anemones from his woodland travels and included a description of this special flower in "Cowper Green," which was written at Helpstone in 1821:

> Anemone's weeping flowers,
> Dyed in winter's snow and rime,
> Constant to their early time,
> White the leaf-strewn ground agen,
> And make each wood a garden then.

English poet Anne Pratt writing in the mid-1800s also remembered the wood anemone in its natural habitat. In "To the Wood Anemone" she wonderfully described a wild wood recalled from her childhood:

> Flowers of the wild wood! your home is there,

'Mid all that is fragrant, all that is fair;
Where the wood-mouse makes his home in the earth;
Where gnat and butterfly have their birth;
Where leaves are dancing over each flower,
Fanning it well in the noontide hour.

Two English women, an aunt, Katherine Bradley, and niece, Edith Cooper, who were "poets and lovers evermore" to each other, wrote under the same pen name, Michael Field. In "Constancy" they compared their relationship to the perennial nature of anemones, also using the flower's shy connotations to explain the private intensity of their love:

I love her with the seasons, with the winds,
As the stars worship, as anemones
Shudder in secret for the sun, as bees
Buzz round an open flower.

—⁂—

The third flower amongst the confusion and interchanging of adonis and anemone is the pulsatilla. *Pulsatilla vulgaris*, with its pale white to vivid violet flowers, is distributed throughout Eurasia. *Pulsatilla halleri* and several subspecies are native from southeastern Europe to the Crimea. Two species are native to central and western North America. The alpine *P. occidentalis*, which is generally found above the timberline in the Cascades, the Olympics, and the Sierra Nevada, has midsummer blooms of creamy white, often with a hint of blue-purple. Some are more vividly flushed purple. The spring-blooming *P. patens* is more widespread in the United States, growing from the Great Plains to the Rockies (and to Siberia on the Asian continent), up to elevations as high as 3700 meters (12,000 feet). In all, about thirty species of pulsatilla are known.

The name *Pulsatilla* is the diminutive of *pulsata*, meaning beaten or driven, from *pulsare*, to beat, suggested by the blowing wind's beating of the flower. But plant taxonomists are a changeable lot; the pasque flower (*Pulsatilla vulgaris*) was once classified and is still frequently found as *Anemone pulsatilla*. The plant is often called the pasque or Easter flower, for it is said to be the first to bloom at Eastertide. In the western United States, the pasque flower, primarily *P. occidentalis*, is sometimes called old man of the mountain because of its hairy leaves and sepals. In the same region, it is grouped with anemones and called wild crocus, a name that adds further nomenclatural confusion.

The exchequer of King Edward I of England kept a household account indicating that a green dye from the pasque flower was used to color Easter eggs. Gilbert White of Selborne, England, referred to the pasque flower in his journals: "April 5, 1769 [Oxford]. Anemone pulsatilla budds. This plant, the pasque flower, which is just emerging and budding for bloom, abounds on the sheep-down just above Streatley in Berks[hire]." According to Old English lore, the pasque or passe flower is known as Dane's blood because it grows only where blood from Danish invaders was "spilt."

These three flowers, adonis, anemone, and pulsatilla, provide such a delicate grace to the garden that they inspired John Parkinson to write effusively, "The sight of them doth enforce an earnest longing desire in the minde of anyone to be a possessour of some of them at the least." He was so enamored that he devoted fourteen pages of *Paradisi in Sole, Paradisus Terrestris* (1629) to the "windeflower and his kindes." Regardless of their proper names botanically and the evolution of their naming, they were well known; by the 1800s, in the Victorian language of flowers each had its distinctive meaning: the adonis suggested painful recollections, anemone represented sickness and expectation, and the pasque flower stated flatly, "You have no claims."

AMARYLLIS

To win the gaze of human eye

Given the popularity and renown of the amaryllis, it is surprising to learn that current taxonomy assigns only one species of this bulbous plant to the genus *Amaryllis*. *Amaryllis belladonna* is native to South Africa where it blooms in late fall. It also blooms outdoors in autumn in California and in the warmer parts of the British Isles. Closely related genera within the Amaryllidaceae, some of which were formerly classified as *Amaryllis*, include *Hippeastrum*, *Rhodophiala*, *Brunsvigia*, and *Crinum*. Several bigeneric hybrids occur, including ×*Amarygia parkeri*, a cross between *Amaryllis belladonna* and *Brunsvigia josephinae*, and ×*Amarcrinum*, a hybrid of *Amaryllis* and *Crinum*.

The amaryllis was given its scientific name, *Amaryllis belladonna*, by Carl Linnaeus, recalling a beautiful shepherdess mentioned in poems by Theocritus, Ovid, Virgil, and Milton. Etymologically, the genus name comes from the Greek verb *amarussein*, meaning to sparkle or shine, and the specific epithet is an Italian word meaning beautiful or fair lady. This linguistic history makes an apparent connection with a cosmetic used by Italian women during the Renaissance. A solution called belladonna was made from a plant extract containing atropine. Women put drops of belladonna into their eyes to dilate the pupils and make their eyes appear to sparkle. The extract, however, came from the deadly nightshade (*Atropa belladonna*), not from the amaryllis. Even so, *Amaryllis belladonna* is often called the belladonna lily, a practice explained in *The Treasury of Botany* (1866) by Thomas Moore and John Lindley: "The name Belladonna Lily was given . . . from the charmingly blended red and white of the perianth, resembling the complexion of a beautiful woman."

Another common name applied to the amaryllis is Jersey lily. It was also a nickname for Lillie Le Breton Langtry, a Victorian English actress born in Jersey who became mistress of Crown Prince Edward VII. However, the name Jersey lily was used for amaryllis as early as the 1600s.

The strikingly beautiful amaryllis is an image that has been used by many writers. Milton wrote in "Lycidas" of being tempted "To sport with Amaryllis in the shade/Or with the tangles of Neaera's hair." Though Milton referred to the shepherdess Amaryllis (and to the nymph Neaera from Homer's *Odyssey*), his image invokes the beauty of both the girl and perhaps the flower. Milton's elegy is arguably the finest in the English language. It was written in 1637 upon the drowning of his university classmate Edward King. Tennyson also conjured up the amaryllis and other flowers in "The Daisy" (1855), a poem of the poet's tender daydream after picking a daisy along the streets of Edinburgh:

> What slender campanili grew
> By bays, the peacock's neck in hue;
> Where, here and there, on sandy beaches
> A milky-bell'd amaryllis blew!

In the language of flowers, the amaryllis suggested haughtiness and pride, owing, it seems, to the dazzling splendor of the amaryllis flower terminating on an elongated scape. Ironically, timidity was also symbolized by the amaryllis, perhaps because it was a trait of the hillside

shepherdesses of ancient Greece. The prideful and disdainful amaryllis was described in an unattributed poem:

> When Amaryllis fair doth show the richness of her fiery glow,
> The modest lily hides her head; the former seems so proudly
>   spread
> To win the gaze of human eye, which soonest brightest things
>   doth spy.
> Yet vainly is the honour won, since hastily her course is run;
> She blossoms, blooms,—she fades,—she dies,—they who
>   admired, now despise.

ARUM

An apoplectic saint in a niche of malachite

Arums are tuberous herbaceous perennials in the Araceae. Distribution of the twenty-six species is centered around the Mediterranean and reaches eastward to the Himalayas. The plants form an elongated spadix, a sort of spike, bearing flowers and a surrounding spathe, a modified leaf. They produce bright orange to red berries and have arrow-shaped leaves. The popularly grown *Arum italicum* sends up its sometimes marbled leaves in autumn and produces flowers in late spring. Other species include *A. concinnatum*, *A. dioscoridis*, and *A. maculatum*. They are closely related to the North American jack-in-the-pulpit (*Arisaema*). Although arum berries are poisonous, the roots were used to produce a fine farinaceous starch that the Irish ate during the potato famine in the mid-1800s; earlier, the paste was used to starch fine Elizabethan lace. The French used the starch to make a cosmetic called cypress powder. However, the stems and leaves contain needle-shaped crystals of calcium oxalate, a powerful skin irritant.

The genus name *Arum* is from the Greek *aron*, the classical name for the plant. It is related to an old Arabic word *ar*, for fire, a reference to the burning or bitter taste of the leaves. The elegant and formal look of the upright spadix and surrounding spathe has resulted in a quaint and picturesque common name, lords and ladies, which most frequently refers to *A. maculatum*. Cuckoo-pint, cuckoo-pintle, and priest's pintle are common names used in some British counties. The word pintle is a colloquialism for penis; thus, young men would place a cuckoo-pintle in their shoe when going out for the evening, reciting, "I place you in my shoe and let all girls be drawn to you." In a more refined sense, arum suggested ardor in the language of flowers.

The common name arum lily is sometimes applied to *Zantedeschia aethiopica*, the calla lily or florist's arum. Though it is a member of the Araceae, this plant is not a true arum, lily, or calla.

In German lore, wherever the arum or *Aronswurzel*, for Aaron's rod, grew and flourished, wood spirits would rejoice. The name Aaron's rod comes from a folktale related to the biblical story of the Israelites' journey to the promised land. Among the belongings that they carried from Egypt was a long staff used to transport heavy bunches of grapes that they picked for food along the way. The staff was called Aaron's rod after the brother of Moses. At one point the men carrying the Aaron's rod grew tired and decided to rest, sticking the staff into the ground. Presently, from that spot an arum sprang up as a sign of the fertility and abundance of the land.

Stories of plant lore say that farmers would judge the size of their forthcoming crops by the size of the spadix on the arum each spring: a large spadix suggested a bountiful crop. And bears awaking from hibernation eat the leaves and roots of arum to provide strength, vitality, and apparently necessary flatulence. John Gerard described the bears' practice in his *Herbal or General History of Plants* (1633 revision by Thomas Johnson):

> Bears after they have lien in their dens forty dayes without
> any manner of sustenance, but what they get with licking
> and sucking their own feet, do as soone as they come forth
> eate the herbe Cuckow-pint, through the windie nature
> whereof the hungry gut is opened and made fit againe to
> receive sustenance: for by abstaining from food so long a

time, the gut is shrunke or drawne so close together, that
in a manner it is quite shute up.

John Clare mentioned arum in two poems. Despite spending part
of his writing life in an insane asylum, he was able to idealize rural
England with his imagery, as he did in "On the Sight of Spring" (1821):

> How sweet it us'd to be, when April first
> Unclos'd the arum-leaves and into view
> Its ear-like spindling flowers their cases burst,
> Beting'd with yellowish white or lushy hue.

And in "Recollections After a Ramble" (1821), when the "rosy day was
sweet and young,"

> Some went searching by the wood,
> Peeping 'neath the weaving thorn,
> Where the pouch-lipp'd cuckoo-bud
> From its snug retreat was torn.

"On Returning to England" is by Alfred Austin, poet laureate despite being much parodied and derided because of his ardent impe-
rialist views and waspish criticisms. He included a stanza of childhood memory, mentioning the arum as cuckoo-pint:

> The children loitering home from school,
> Their hands and pinafores all full
> Of cuckoo-pint and bluebell spike,
> Gathered in dingle, dell and dyke.

Perhaps the most picturesque and poetic image of the arum was
conjured by Thomas Hardy in *Far from the Madding Crowd* (1874). In one
particular scene among the many rural settings described, it is the first
day of June and the sheep-shearing season has culminated: "Flossy
catkins of the later kinds, fern-sprouts like bishops' croziers, the
square-headed moschatel, the odd cuckoo-pint,—like an apoplectic

saint in a niche of malachite,—snow-white ladies'-smocks, the tooth-wort, approximating to human flesh, the enchanter's night-shade, and the black-petaled doleful-bells, were among the quainter objects of the vegetable world in and about Weatherbury at this teeming time."

Hardy also referred to the arum in *Tess of the D'Urbervilles* (1891). Tess, while working as a milkmaid, was wooed by Angel Clare, the son of a clergyman: "[Angel Clare] observed [Tess's] dejection one day, when he had casually mentioned something to her about pastoral life in ancient Greece. She was gathering the buds called 'lords and ladies' from the bank while he spoke."

ASPHODEL

My regrets follow you to the grave

The twelve species that make up the genus *Asphodelus* are distributed from the Mediterranean to the Himalayas. They are members of the Asphodelaceae, formerly in the Liliaceae. The white asphodel of southern Europe is *A. albus*, and the short-stemmed, pink-flowered asphodel of Algeria and Morocco is *A. acaulis*. The yellow asphodel, formerly *A. luteus*, is now classified as *Asphodeline lutea*. The name *Asphodelus* is the classic Greek name for *A. ramosus* and other true asphodels. Several related plants can be misidentified as asphodels based on former botanical classification or their similar appearance, as in the case of the false asphodel of the American Southeast, *Tofieldia racemosa*.

The white flowers of the asphodel have been a symbol of death for centuries. The Greeks planted them near tombs in the belief that the roots would nourish the shades, spirits that had departed their deceased physical bodies. Part of Greek mythology supposed that beyond the River Acheron, one of five rivers that flowed in the underworld, there were great fields of asphodel where the shades wandered. The plant is the flower of the dead in classical literature, a role that arises from its part in the story of Persephone, the goddess of crop renewal. Once, as she walked the hills of Sicily, she found a beautiful asphodel growing in the open fields. When she went to pluck it, the ground opened and Hades appeared and seized her. He took her away in a golden chariot to be his bride in hell. The asphodel became her symbol as queen of the infernal regions. In keeping with the myth, the language of flowers recorded that asphodels said, "My regrets follow you to the grave."

Shakespeare alluded to the kidnapping of Persephone, and the same story was retold by poet Jean Ingelow. Both writers named the daffodil as the flower Persephone picked. The names daffodil and asphodel appear to have a common origin, the former being an unexplained variation of the Middle English word *affodile*, which itself evolved from the medieval Latin *affodillus*. Some philologists suggest that the name daffodil arose from the Dutch or Flemish *de affodile*. The variations daffadilly and daffadowndilly appear in Edmund Spenser's *Shepheardes Calender* (1579). To add to the confusion, at least one writer has called the asphodel "a narcissus of marvellous beauty." But in Ingelow's poem "Persephone" (1862), it seems probable that the word daffodil is a synonym for asphodel:

> She stepped upon Sicilian grass
> Demeter's daughter fresh and fair,
> A child of light, a radiant lass,
> And gamesome as the morning air.
> The daffodils were fair to see,
> They nodded lightly on the lea.
>
> . . . .
>
> "O light, O light!," she cries, "farewell;
> The coal-black horses wait for me.
> O shade of shades, where I must dwell,
> Demeter, mother, far from thee!
> Oh, fated doom that I fulfill!
> Oh, fateful flower beside the rill!
> The daffodil, the daffodil!"

St. Cecilia was a Roman maiden martyred in the second or third century. Alexander Pope wrote "Ode for Music on St. Cecilia's Day"

(1713) in her honor and in appreciation of the concerts and recitals that have often been given on her feast day, 22 November, since she was proclaimed the patron saint of music in Rome in 1584. In the poem, Orpheus, possessor of magic musical powers, invokes the deities of the lower world to bring his wife, Eurydice, back to him. *Asphodeline lutea* is probably the yellow asphodel mentioned:

> By the streams that ever flow,
> By the fragrant winds that blow
> O'er th' Elysian flow'rs;
> By those happy souls who dwell
> In yellow meads of Asphodel,
> Or Amaranthine bow'rs;
> By the hero's armed shades,
> Glitt'ring thro' the gloomy glades;
> By the youths that dy'd for love,
> Wand'ring in the myrtle grove,
> Restore, restore Eurydice to life:
> Oh take the husband, or return the wife!

Elizabeth Barrett Browning in *Sonnets from the Portuguese* (1850), sonnet 27, used the asphodel in its capacity as the symbol of the Greek shades in their fields across the River Acheron. She wrote of the effect love had on her, comparing her life before to what it is now, just as the shades must have compared their lives tethered to physical bodies with their later spiritual existence:

> My own, my own,
> Who camest to me when the world was gone,
> And I who looked for only God, found thee!
> I find thee; I am safe, and strong, and glad.
> As one who stands in dewless asphodel
> Looks backward on the tedious time he had

In the upper life,—so I, with bosom—swell,
Make witness, here, between the good and bad,
That Love, as strong as Death, retrieves as well.

In book 9 of *Paradise Lost* (1667), Milton, too, used asphodels to enhance the image of love. Adam and Eve found "their fill of love and love's disport" in the embowered woods of Eden on a carpet of asphodels: "[Adam] led her, nothing loth; flowers were the couch,/Pansies, and violets, and asphodel,/And hyacinth—Earth's freshest, softest lap."

ASTER

Astraea's tears of star dust

Asters are ubiquitous. These herbaceous perennials or shrubs consist of at least 250 species distributed around the globe. Those well known in the garden include *Aster novae-angliae*, a vigorous aster of New England that may grow to 2 meters (7 feet) tall; *A. carolinianus*, the climbing Carolina aster of the southeastern United States; pink, blue, and lilac cultivars of *A. ericoides*; and hybrids of *A. novi-belgii*, the Michaelmas daisy. Another gardenworthy plant is *A. alpinus* from the Alps and the Pyrenees. The majority of garden asters are hybrids that include at least *A. novae-angliae* and *A. novi-belgii* in their genetic makeup.

With botanical nomenclature ever changing and with numerous common names in use from region to region, the name aster is often used to refer to other genera in the Compositae (Asteraceae). For example, the once popular China aster is now classified as *Callistephus chinensis*. In Victorian England, the aster most commonly grown and cited was the China aster, an annual with a variety of seed races and hybrids and a rainbow of colors. It is this wide selection of vivid colors that suggested the aster as a symbol for variety in the language of flowers. Some popular species of aster that originated in North America were exported to Europe; for example, John Tradescant, the younger, who made three plant-collecting trips to Virginia, carried *A. tradescantii* and other asters to London in 1637. Generally much improved hybrids of those transplanted asters have since been reintroduced to the United States and Canada.

Asters are named from the Latin word for star because of their star-shaped flower heads. The astronomical meaning of the word is now obsolete in English, but as late as 1632 John Florio described the "revolutions . . . and carrols [dances] of the asters and planets" about the sun. By the early 1700s the name aster was in use for the plant, along with the other common names star plant, starwort, and occasionally sparewort. The etymological links survive in modern English in words such as asterisk, for little star, and disaster, relating to evil influence, as in ill-starred or "against the stars."

The name aster also has roots in Greek mythology. During the Golden Age, the time when the gods created the earth and the creatures upon it, the time before evil and hardship entered the lives of men, Astraea dwelt on the earth as the goddess of innocence. When sin began to prevail she left and was metamorphosed into the constellation Virgo. Zeus, angered by the sinful conditions on the earth, created a flood that covered all except the peak of Mt. Parnassus, to which fled the last surviving people, Deucalian and his wife, Pyrrha. As the waters receded and the couple became lost and lonely in their muddy surroundings, Astraea created starlight to guide them. Her tears fell as stardust that covered the earth and changed into star-shaped flowers, the asters or starworts.

Aster amellus or the Italian starwort is the original plant dedicated to the archangel Michael, the highest in rank of the seven archangels and leader of the celestial armies, as this unattributed couplet explains: "The Michaelmas daisy, among dede weeds/Blooms for St. Michael's valorous deeds." The English chose instead to dedicate to St. Michael their native English starwort, *A. tripolium*. Nowadays, the Michaelmas daisy is generally *A. novi-belgii* or hybrids and cultivars derived from it after its introduction to Europe from America in the 1600s. Michaelmas, the festival of St. Michael and All Angels, occurs on 29 September, one of the four quarter days in England. The festival also lends its name to the fall semester of English universities, when the plant is at its peak bloom. The Michaelmas daisy has become a well-known repatriate in American gardens.

The virtues of "starre-wort" were described by John Parkinson "to bee good for the biting of a mad dogge, the greene herbe being beaten with old hogs grease, and applyed." John Gerard noted that the sea starwort grows along the coast but that if it is brought into a garden it flourishes. He called it tripolium because "It is reported by men of great fame and learning [he was referring to Dioscorides, a Greek physician of the first century], that this plant was called Tripolium because it doth change the colours of his floures thrice a day." Pliny the Elder recommended a tea of aster to assuage snake bite and an aster amulet for easing sciatica pain. Virgil said that the roots of *Aster amellus* were a physic for curing ailing bees and that wreaths of it were used to adorn the altars of the gods. In his *Georgics* (37–30 BCE, translated by T. C. Williams in 1915), a didactic poem on agriculture and farm life, Virgil described in detail the wild Italian starwort:

> There is a useful flower
> Growing in the meadows, which the country folk
> Call star-wort, not a blossom hard to find,
> For its large cluster lifts itself in air
> Out of one root; its central orb is gold
> But it wears petals in a numerous ring
> Of glossy purplish blue; 'tis often laid
> In twisted garlands at some holy shrine.
> Bitter its taste; the shepherds gather it
> In valley-pastures where the winding streams
> Of Mella flow. The roots of this, steeped well
> In hot, high-flavored wine, thou may'st set down
> At the hive door in baskets heaping full.

The aster, or some closely related daisy, is the flower that Goethe chose to use in his tragic drama *Faust* (1808). In the famous garden scene, repeated in Charles Gounod's 1859 opera of the same name, Margaret plucks off the florets of an aster one by one, as an oracle, and on the last petal realizes, "He loves me!" And Faust responds, "Yes. Believe this flower at your feet. Let it be a sign from heaven to your heart."

William Cullen Bryant included aster in "My Autumn Walk" (1864) among the plants he saw in an October field as his thoughts drifted toward the Civil War:

> The golden rod is leaning,
> And the purple aster waves
> In a breeze from the land of battles
> A breath from the land of graves.

Mary Delany wrote in her *Autobiography and Correspondence* (1779–1788, published 1861) of "A little pale purple Aster with a yellow thrum." John M. Jephson noted in *Narrative of a Walking Tour in Brittany* (1859) that there was "A fine show of aster in full bloom." And Bliss Carman, describing an autumn ramble in "A Vagabond Song," wrote of the gray fruiting heads:

> The scarlet of the maples can shake me like a cry
> Of bugles going by.
> And my lonely spirit thrills
> To see the frosty asters like smoke upon the hills.

BEGONIA

Beware. I am fanciful

About 900 species of *Begonia* grow widely through both tropical and mild temperate regions. They are perennial members of the Begoniaceae and may be herbaceous, shrubby, or climbing. Many species are shade lovers, and their leaves are generally asymmetric and waxy. Because most of the species, cultivars, and hybrids are not hardy, they are typically grown as houseplants or as bedding plants for summer outdoor color. *Begonia ×erythrophylla* is the beefsteak, kidney begonia, or wax-leaf begonia; it is a cross between *B. hydrocotylifolia* and *B. manicata. Begonia sutherlandii* is a popular tuberous perennial with annual stems and delightful, small, orange-salmon flowers. *Begonia grandis* subsp. *evansiana*, the "hardy" begonia (Zone 6), has forms with both white and pink flowers. *Begonia rex* is famed for its colorful metallic-looking leaves with silver-white markings.

The botanist and monk Charles Plumier first discovered begonias in Mexico in 1690. He named the plants after Michael Bégon, a government official assigned to Santo Domingo under Louis XIV and later a governor of French Canada.

Bégon was a patron of botany and the sciences, but he probably never knew the plant Plumier named for him. Begonias did not arrive in England until the mid-1700s and did not become popular until nearly a century later, when more species, including *Begonia boliviensis*, arrived in London at the Royal Botanic Gardens, Kew, from tropical parts of America and from both the East and West Indies. *Begonia grandis* was a popular addition from China. These new arrivals were displayed at flower shows in London and Paris and became popular Victorian houseplants, particularly the hybrid forms.

In the language of flowers, much in use in Victorian England, the begonia came to symbolize dark thoughts. The plant could also send the message, "Beware. I am fanciful"—an allusion to the showy, feathery hybrid forms.

Ephraim Chambers was the first to write of begonias. In his 1751 *Cyclopedia* (the first English-language encyclopedia) he described "The great purple Begonia with auriculated leaves." Mary E. Braddon, poet, actress, and novelist known for seventy-five lurid crime novels, wrote in *Asphodel* (1881) that "All the tribe of begonias, and house-leeks, [are] newly bedded out." And *The Pall Mall Gazette*, an upper-class London newspaper that published news, commentary, and literature from 1865 to the 1920s, noted in September 1883, "The well-known Begonias and Fuchsias . . . have withstood the late storms better than any of their rarer rivals."

Mrs. C. W. Earle, the pen name of Maria Theresa Villiers Earle, published two diaries prompted by ruminations on her garden through the seasons. She gardened at Woodlands in Cobham, and her niece, Emily Lytton, was married to the garden designer Edwin Lutyens who collaborated with Gertrude Jekyll and through whom Earle and Jekyll became friends. Earle discussed begonias in *More Pot-Pourri from a Surrey Garden* (1899), the second of her two books:

> JANUARY 6, 1899. A little winter-flowering Begonia, called *Gloire de Lorraine*, has lately come into fashion. What a term

for a flower! But it is true, and plants of this Begonia make a charming table decoration at a time of year when flowers are scarce. They look best growing in pots. Roman hyacinths in glasses could be placed between, and pink shades used for the candles; or, for a small table, one plant in the middle would be enough. The colour, the growth, the shape of the leaves, all make it charming. I do not know if it is difficult to grow, as I have only lately bought a plant.

After only a month of experience with begonias, she continued to praise them:

JANUARY 27, 1899. I have on my flower table a shrubby Begonia in a pot with small, pointed, spotty leaves and hanging white flowers. They are easily reared from seed, and I do think they grow so beautifully and can be pruned into such lovely shapes! They are far more beautiful than those great, flat, floppy, opulent, tuberous-rooted ones that flower in the summer. The parent of my plant (mossy green leaves, spotted silvery-white) must have been called *Begonia alba picta* [*B. albopicta*].

BELLFLOWER

The auld~man's~bells

The genus *Campanula*, the bellflower, consists of more than 300 species distributed throughout the Northern Hemisphere, particularly in the Mediterranean region. Bellflowers are in the Campanulaceae and range from the diminutive, ground-hugging, 3-cm-tall (about 1-inch) *C. calaminthifolia* to the stately meter-tall (3-foot) *C. persicifolia*, the peach-leaf bellflower of border fame. Among the common campanulas are fairies' thimbles, *C. cochleariifolia*; Coventry bells or throatwort, *C. trachelium*; and the famous bluebells or harebells of Scotland, *C. rotundifolia*, whose distribution is circumpolar in the Northern Hemisphere. Plants of the genera *Hyacinthus* and *Hyacinthoides* are also called bluebells, but they are in the botanically unrelated Liliaceae (or the smaller, more recently defined Hyacinthaceae) (see the chapter "Hyacinth"). *Campanula poscharskyana* and its many cultivars are available in the trade. In North America, native campanulas include *C. divaricata*, the fairy bluebell with diminutive flowers from the eastern United States, and *C. piperi*, an evergreen from the western United States. One species, *C. rapunculus*, the native rampion of Europe, was formerly cultivated for its tubers that were eaten uncooked or boiled and its leafy tops that were used for salads.

The name campanula is the diminutive of *campana*, Latin for bells, and means little bell, referring to the shape of the flower. Perhaps the most well-known campanula is Canterbury bells, now applied to *Campanula medium*, a biennial. The source of the common name is an English legend in which three wicked young men were turned into swans by a priest and were forced to fly endlessly for a thousand and one years. On one flight over Canterbury, they heard the nearby ringing of the Christian church bells. They were awed by the sound and the spell was broken. They fell to earth at Canterbury where St. Augustine found them lost and bewildered and led them into the church. As they went inside, little bell-shaped flowers (probably *C. trachelium*) sprang up from the ground where they stepped. These flowers were dedicated to St. Augustine and later to St. Thomas à Becket, murdered in the cathedral at Canterbury in 1170. Another version of the naming of these bell-shaped flowers tells of the pilgrims who traveled to Canterbury carrying poles with horsebells, cowbells, or small handbells so that the jangling and ringing would signify that they were harmlessly on a pilgrimage. The Canterbury bells were named for their resemblance to the bells the pilgrims carried while wending their way to pray at St. Thomas's shrine. Common usage over time shortened the name to belfloure, and now bellflower is primarily used for the garden forms of *Campanula*.

The bluebells of Scotland have made their way into legend as well. No one would dream of picking bluebells, for they are the auld-man's-bell, the bell of the resident ghost of the graveyard who comes out on stormy nights. If one hears his sweet ringing of the bluebells above the thunder, death will come within a fortnight. Hence, the flowers are to remain unmolested, according to legend.

In literature, bellflowers made their appearance in the late sixteenth century when Henry Lyte wrote in his translation of Rembert Dodoens's *Niewe Herball or Historie of Plantes* (1578): "In English they be called Bel-floures and of some Canterbury Belles." Shakespeare, whose literary garden is full of flowers, noted the harebell, *Campanula rotundifolia*, in act 4 of *Cymbeline* (1623). Carrying the lifeless Imogen in his arms, Arviragus intones,

With fairest flowers
Whilst summer lasts, and I live here, Fidele,
I'll sweeten thy sad grave. Thou shalt not lack
The flower that's like thy face, pale primrose; nor
The azured hare-bell, like thy veins; no, nor
The leaf of eglantine, which, not to slander,
Out-sweetened not thy breath.

In book 6 of *The Prelude* (1805) William Wordsworth described the river and old stone towers of Yorkshire, at the same time tracing the path of his own life. On a summer's day, wandering along the margin of a stream, thinking of his poetic vocation and his humility, he heard the harebells:

Catching from tufts of grass and hare-bell flowers
Their faintest whisper to the passing breeze,
Given out, while mid-day heat oppressed the plains.

Sir Walter Scott seemed particularly enamored of the bellflower. In "Rokeby," a poem in six cantos published in 1813, the complicated plot takes place near Greta Bridge in Yorkshire after the 1644 battle of Marston Moor. The ruffian Bertram Risingham, hunted by Redmond O'Neale for plotting to murder Philip of Risingham, hides in the woods along the River Greta in a carpet of throatwort, which because of the location and habitat is more probably *Campanula rotundifolia*:

He laid him down,
Where purple heath profusely strown,
And purple throatwort, with its azure bell,
And moss and thyme, his cushion swell.

In *The Lady of the Lake* (1810) Scott used the harebell to introduce the heroine, Ellen Douglas, showing her natural refinement through her effect on the flower:

What though no rule of courtly grace

To measured mood had train'd her pace;
A foot more light, a step more true,
Ne'er from the heathflower dash'd the dew;
E'en the slight harebell raised its head,
Elastic from her airy tread.

Later in the poem, Ellen picks the flower in recognition of its simplicity:

"For me,"—she stoop'd, and looking, round,
Pluck'd a blue harebell from the ground,—
"For me, whose memory scarce conveys
An image of more splendid days,
This little flower that loves the lea,
May well my simple emblem be."

In "Sonnet Written In November," John Clare described the late-blooming bellflowers that he saw along his countryside rambles: "And in the moss-clad vale, and wood-bank wild/Have cropt the little bell-flowers, pearly blue,/That trembling peep beneath the shelt'ring bud behind."

Englishman Robert Browning, whose early years as a poet were highly unsuccessful, spent fifteen years in Italy largely because of his predilection for the Italian Renaissance and its subjects. During his Italian residency he wrote "A Toccata of Galuppi's" (1855), a poem detailing the thoughts of an Englishman as he plays a toccata of the eighteenth-century composer Baldassare Galuppi; his reflections roam to St. Mark's Cathedral, doges, the Grand Canal, and "Venice and her people, merely born to bloom and drop." At one point his thoughts rest on a woman, whose beauty he likens to that of the bellflower: "Was a lady such a lady, cheeks so round and lips so red,—/On her neck the small face buoyant, like a bell-flower on its bed,/O'er the breast's superb abundance where a man might base his head?"

Thomas Edward Brown, a late-nineteenth-century poet from the

Isle of Man off the northeast coast of Ireland, told about Manx life in "A Fable: For Henricus D., Esq. Jun." (1900). In this passage, he related the mythical heritage of his home:

> In the old, old times
> The harebells had their chimes,
> · I can tell you, and could sing out loud and brave;
> But Queen Titania said
> That they quite confused her head.

In the passionate prose of *Wuthering Heights* (1847), Emily Brontë recalled the bluebells or harebells native to the moors of Yorkshire. The young Cathy Linton, niece of Heathcliff, is on an outing with her nurse one October day, and both of them pause to study, high up on a rough bank, the last trembling bluebell left from the summer:

> "Look, miss!" [the nurse] exclaimed, pointing to a nook under the roots of one twisted tree; "Winter is not here yet. There's a little flower up yonder—the last bud from the multitude of bluebells that clouded those turf steps in July with a lilac mist. Will you clamber up and pluck it to show to papa?"
> Cathy stared a long time at the lonely blossom trembling in its earthy shelter, and replied at length, "No, I'll not touch it. But it looks melancholy, does it not, Ellen?"

And, in the final lines of Brontë's masterwork, Mr. Lockwood, one of the narrators of the story and a visitor at Wuthering Heights, locates the headstones of Heathcliff and Catherine next to the moor near the old kirk:

> I lingered round them, under that benign sky: watched the moths fluttering among the heath and hare-bells; listened to the soft wind breathing through the grass; and wondered

how any one could ever imagine unquiet slumbers for the sleepers in that quiet earth.

English Victorian poet Constance Naden in "Poet and Botanist" (1894) showed the creative miracle of the poet's art by comparing it to the flowering and fruiting of the bellflower:

> Fair are the bells of this bright-flowering weed;
> Nectar and pollen treasuries, where grope
> Innocent thieves; the Poet lets them ope
> And bloom, and wither, leaving fruit and seed
> To ripen.

Even those poets whose names are lost to us today were charmed by the bellflower. One described a boyish trick of putting a glowworm in the flower of a Canterbury bell to make a small lantern:

> When glow-worm, found in lanes remote
> Is murdered for his shining coat,
> And put in flowers that Nature weaves,
> With hollow shape and silken leaves,
> Such as the Canterbury-bell
> Serving for lamp or lantern well.

Another poet lovingly described the sight of the blooms in "My Ain Bluebell, My Bonnie Bluebell":

> Light Harebell, there thou art
> Making a lovely part
> Of all the splendour of the days gone by,
> Waving, if but a breeze
> Pant through the distant trees,
> That on the hill-top grow broad-branched and high.

BROOM

To lie at rest among it

Two genera of plants are commonly called broom: *Genista* and *Cytisus*. *Genista* is a shrubby plant that bears yellow, pea-like flowers and tiny leaflets. It is comprised of nearly ninety species in the Leguminosae and is distributed from southern Europe through the Mediterranean to western Asia. Brooms include *G. tinctoria*, the dyer's greenweed; *G. hispanica*, the Spanish gorse; *G. anglica*, the needle furze; and *G. pilosa*, the woadwaxen. *Cytisus scoparius*, the common broom or Scotch broom, is probably the plant most widely referred to as broom. Its specific name means broom-like. Sharing visual similarities, common names, and botanical family with the above genera is *Ulex europaeus*, called gorse or furze in some regions of Europe. *Genista* is the classical Latin name selected by Linnaeus for the broom shrub. *Ulex* was the classical Latin name for gorse, which Pliny applied to heathers. *Cytisus* is a name derived from the Greek *kytisos*, which was applied generally to woody legumes.

For centuries broom has been a highly useful plant. Some species yield a yellow dye and others have been the source of perfume. Some species are good fodder for animals, and the root fibers have made cordage for ship rigging. It was also used as a poor man's thatching on outhouses and sheds. And as the name suggests, it has been used as a household besom, or broom, because of its flexible twigs.

The English royal line of Plantagenet took its name from the broom's medieval name *planta genesta*. Geoffrey IV, duke of Anjou and father of Henry II, first gathered the yellow flower on a rocky pathway, saying that this golden plant should "ever be my cognizance rooted firmly amid rocks and yet upholding that which is ready to fall." From then on he wore a sprig of broom on his helmet in battle. Geoffrey's heirs assumed the surname Plantagenet (the French for broom is *genêt*), and from him were descended fourteen monarchs of England from Henry II through Richard III, reigning from 1154 to 1485. In time the flower was embodied on the family's coat of arms. An anonymous poem tells the story of the yellow broom flower and its association with the Plantagenets:

> Time was when thy golden chain of flowers
> Was linked, the warrior's brow to bind;
> When reared in the shelter of royal bowers,
> Thy wreath with a kingly coronal twined.
>
> The chieftain who bore thee high in his crest,
> And bequeathed to his race thy simple name,
> Long ages past has sunk to his rest,
> And only survives in the rolls of fame.
>
> Though a feeble thing that Nature forms,
> A frail and perishing flower art thou:
> Yet thy race has survived a thousand storms,
> That have made the monarch and warrior bow.
>
> The storied urn may be crumbled to dust,
> And time may the marble bust deface;
> But thou wilt be faithful and firm to thy trust,
> The memorial flower of a princely race.

In yet another brush with royalty, the broom was selected by Louis IX, king of France from 1226 to 1270, as a badge for a new order of knighthood to mark the coronation of Queen Marguerite in 1234. The knights of the order under St. Louis, as he became known, wore a bouquet of the yellow blossoms of genêt entwined with white fleurs-de-lis. Nobles associated with the order wore coats with broom embroidered on front and back with the inscription *Exaltat humiles* ("He exalteth the humble"). This slogan may be behind part of the meaning given to broom in the language of flowers: humility and neatness.

The yellow broom turns into gold in a ballad in which a young Breton sings of his sweet love:

> If on my door-sill up should come
> Golden glowers for furze and broom
> Till my court were gold piled high,
> Little I'd reck but she were by.

The Scottish poet and scholar of the Orient John Leyden was inspired equally by the ballads of his native land and by tales from the Far East. In his poem "Noontide," he reclines beneath a canopy of aspen overlooking a blossomed lea and observes:

> [I] wander mid the dark-green fields of broom,
> When peers in scattered tufts the yellow broom;
> Or trace the path with tangling furze o'er run,
> When bursting seed-bells crackle in the sun.

English Victorian poet Mary Botham Howitt remembered glittering bushes of broom at her mother's door and all the way down in the glen. She reminisced in "The Broom-Flower,"

> O the broom, the yellow broom!
> The ancient poet sung it,
> And dear it is on summer days
> To lie at rest among it.

Robert Burns wove a bower of yellow broom for his sweetheart in "A Song":

Their groves o' sweet myrtle let foreign lands reckon,
Where bright beaming summers exalt their perfume;
Far dearer to me yon lone glen o' green bracken
Wi' the burn stealing under the lang yellow Broom.

Far dearer to me are young humble Broom bowers,
Where the bluebell and gowan lurk lowly unseen;
For there, lightly tripping amang the sweet flowers,
A-listening the linnet, oft wanders my Jean.

Finally, in "Lessons from the Gorse" Elizabeth Barrett Browning sees strength in the mountain gorse and questions what secret is at its heart:

Mountain Gorses, ever golden!
Cankered not the whole year long!
Do you teach us to be strong,
Howsoever picked and holden
Like your thorny blooms, and so
Trodden on by rain and snow
Up the hill-side of this life, as bleak as where ye grow?

Like many other plants, broom plays a part in superstition. Those from the counties of Suffolk and Surrey considered it unlucky to bring into the house while in bloom: "Sweep the house with blossomed broom in May,/And you sweep the luck of the house away." The plant is part of biblical legend as well. When Mary and Joseph fled to Egypt with the infant Jesus, the broom plant under which they took shelter crackled its pod and nearly betrayed their presence. Another legend tells that when Jesus knelt in the Garden of Gethsemane, he stepped on the broom and rustled its pods, giving away his presence to Judas. And finally, let us not forget that witches ride on brooms.

BUTTERCUP

Like golden sunbeams, brightest lustre shed

The buttercup is a well-known flower and is one of the first we recognized as children. Appropriately, in the language of flowers the buttercup signifies childishness and child-like cheerfulness. Those of us with rural backgrounds remember buttercups gleaming along ditches and roadsides, in fields, and in meadows. According to the plant lore of rural England, the blaze of golden buttercups in meadows in which cows grazed was responsible for the rich color of butter.

Botanists call the buttercup *Ranunculus*, which is the Latin diminutive of *rana*, meaning frog, a name chosen by Linnaeus because many of the species grow in wet places inhabited by the amphibians. The 400–600 species of buttercups are scattered worldwide. Some common species are *R. acris*, the meadow buttercup; *R. repens*, creeping buttercup; and *R. aquatilis*, the water crowsfoot of Europe. *Ranunculus bulbosus* has a swollen rootstock, which gave it the old name of St. Anthony's turnip. Although most buttercup flowers are golden yellow, some species' petals are red, orange, or white. *Ranunculus asiaticus* has flowers whose colors range from red to purple; and some improved strains of it, 'Color Carnival' and 'Picotee', exhibit bright oranges and pinks on double flower heads. The buttercup was a popular pot plant in the Victorian era.

The buttercup is known by various

names, including pilewort, crow's foot or crowfoot, and the lesser celandine, names that have been captured in poetry and prose. The two names pilewort and lesser celandine belong to the same plant, *Ranunculus ficaria*, but it is botanically distinct from the celandine, or swallowwort, which belongs to the genus *Chelidonium*, and from the celandine poppy, *Stylophorum diphyllum*, which is native to the eastern United States. The English word celandine is from the Latin *chelidonia*, meaning "of the swallow," which prompted one anonymous herbalist to write that the celandine is so called because "it beginneth to spring and to flower at the coming of the swallows, and withers at their returne." Legend claimed that swallows fed celandine to their young to promote the development of their eyesight. The celandine, at more than a meter (3 feet) tall with yellow-orange petals, looks vaguely like a grand version of *R. ficaria*; thus arose the application of the name lesser celandine. Shakespeare called the buttercup the cuckoobud in "The Song of Spring," which closes *Love's Labour's Lost* (1598): "And cuckoobuds of yellow hue/Do paint the meadows with delight" (his description suggests *Caltha palustris*, however). In his eighteenth-century *Dictionary of the English Language*, Samuel Johnson wrote simply of "butterflower—a flower with which the fields abound in the month of May." Adding to the confusion is the common name kingcups, which is applied to both *Ranunculus* and *Caltha palustris*, the marsh marigold. Both belong to the Ranunculaceae, inhabit marshy areas, and are yellow.

Because of the buttercup's popularity, it often appears in lore and myth. Crowfoot was the traditional decoration used in the Old English church on 3 May, the day to commemorate the invention of the cross. The pilewort, *Ranunculus ficaria*, was a treatment for painful piles, according to herbalists who followed the Doctrine of Signatures, a seventeenth-century explanation of the herbal uses of a multitude of plants, assigning their healing powers based on visual resemblances between plants and parts of the human body.

According to legend, a young poet named Ranunculus had a warm, melodious voice that charmed everyone who heard him, including the nymphs. He often wore gorgeous attire of yellow and green flowing silks. One day while singing he became so entranced with his own sweet voice that he expired in joyous ecstasy. To honor him the gods transformed him into a buttercup that bears his name. René Rapin told the story of this charmer's transformation:

> Ranunculus, who with melodious Strains
> Once charmed the ravished Nymphs on Libyan plains,
> Now boasts through verdant Fields his rich Attire,
> Whose love-sick Look betrays a secret Fire;
> Himself his song beguiled and seized his Mind
> With pleasing Flames for other Hearts designed.

Literary references to the buttercup, in all its various names, abound. In Shakespeare's *Hamlet* (1601), Queen Gertrude describes to Laertes the drowning of his sister Ophelia, who wore flowers in her madness: "There with fantastic garlands did she make/Of crowflowers, nettles, daisies, and long purples."

Wordsworth also praised the lesser celandine in "To the Small Celandine" (1802):

> Long as there are violets,
> They will have a place in story:
> There's a flower that shall be mine,
> 'Tis the little celandine.
>
> . . . .
>
> They have done as worldlings do,
> Taken praise that should be thine,
> Little, humble Celandine.

Wordsworth, who lived most of his life in the Lake Country of England, further described the flower's spring arrival in "The Small Celandine" (1804):

There is a Flower, the lesser Celandine,
That shrinks, like many more, from cold and rain;
And, the first moment that the sun may shine,
Bright as the sun himself, 'tis out again!

John Gay, English poet and playwright born in the late seventeenth century, wrote of butterflowers at a funeral in "Friday, or the Dirge" in "The Shepherd's Week":

To show their love, the neighbours far and near,
Follow'd with wistful look the damsel's bier.
Sprigg'd rosemary the lads and lasses bore,
While dismally the Parson walk'd before.
Upon her grave the rosemary they threw,
The daisy, butterflower and endive blue.

By the late eighteenth century, butterflowers began to be called butter-cups. The "peasant poet" of Northamptonshire, John Clare, often wrote fondly of the buttercup in his descriptions of rural life in England. In "The Village Minstrel" (1821) he said,

There once were summers, when the crowflower buds,
Like golden sunbeams, brightest lustre shed.

. . . .

Before the door, with paths untraced,
The green sward many a beauty graced;
And daisy there, and cowslip, too,
And buttercups of golden hue.

In "The Eternity of Nature" Clare saw the "humble-bee" darting from flower to flower but not to the buttercup:

Who breakfasts, dines, and most divinely sups,
With every flower save golden buttercups,
On whose proud bosoms he will never go,

> But passes by with scarcely "how do ye do?"
> Since in their showy, shining gaudy cells,
> Haply the summer's honey never dwells.

While in confinement in an insane asylum, where he spent his last two decades, Clare wrote in "Spring Comes" of remembrances of "water-buttercups" down by a pond "when frogs and toads are croaking." He was likely describing a white-flowered, aquatic species, *Ranunculus aquatilis,* the crowsfoot.

Robert Browning in "Home-Thoughts, From Abroad" (1845), which contains the famous lines "Oh to be in England/Now that April's there," also reminisced about buttercups:

> And though the fields look rough with hoary dew,
> All will be gay when noontide wakes anew
> The buttercups, the little children's dower
> —Far brighter than this gaudy melon-flower!

Pear-tree blossoms are the "melon-flower" he mentioned.

Elizabeth of *Elizabeth and Her German Garden* (1898), a book by Mary Annette von Arnim-Schlagenthin, gives a loving tribute to celandine in a garden

> surrounded by cornfields and meadows, and beyond [where] are great stretches of sandy heath and pine forests. . . . How happy I was! I don't remember any time so perfect since the days when I was too little to do lessons and was turned out with sugar on my eleven o'clock bread and butter on to a lawn closely strewn with dandelions and daisies . . . . The celandines in particular delighted me with their clean, happy brightness, so beautifully trim and newly varnished, as though they too had had the painters at work on them. . . . I felt so absolutely happy, and blest, and thankful, and grateful, that I really cannot describe.

This book is largely an autobiographical account of von Arnim-Schlagenthin, who wrote under the pen name Elizabeth or Elizabeth von Arnim. The book focuses on her planting an English garden in the north German countryside as an escape from her husband, a Prussian count, for herself and their three daughters.

Buttercups reminded Eliza Cook of the sweetness of childhood love. In "Buttercups" she looked back with nostalgia:

> 'Tis sweet to love in childhood, when the souls that we
>   bequeath
> Are beautiful in freshness as the coronals we wreathe;
> When we feed the gentle robin, and caress the leaping hound,
> And linger latest on the spot where buttercups are found.

Cook, a Victorian poet whose works are relentlessly sentimental, described the emotional effect of a bouquet in "Buttercups and Daisies":

> I never see a young hand hold
> The starry bunch of white and gold,
> But something warm and fresh will start
> About the region of my heart.

Coventry Patmore, an English poet of the nineteenth century, painted a vivid landscape of green meadows and buttercups in "Amelia":

> And broadcast buttercups at joyful gaze,
> Rending the air with praise,
> Like the six-hundred-thousand-voiced shout
> Of Jacob camped in Midian put to rout.

In an era when exotic tender plants from tropical parts of South America were being introduced in greenhouses and were obsessing gardeners, the Victorian horticulturist Shirley Hibberd wrote practical gardening information. In *The Amateur's Flower Garden* (1871), he had these understated observations on the lesser celandine or ficaria:

This sweet little early-flowering British weed is most valuable for damp shady spots, where few other plants will grow, its bright green leaves and golden flowers being most welcome in the early spring. We have seen patches in most unpromising spots in dark, damp sour town gardens, and therefore it must have a place in this selection. All the varieties spread rapidly if the position suits them.

Students of musical theater cannot overlook the anthropomorphology in Gilbert and Sullivan's *HMS Pinafore* (1878). It is easy to see the connotations of childlike cheerfulness borrowed from the language of flowers for naming Buttercup, the female character who is traditionally dressed in bright yellow and portrayed as a plump, good-natured creature. She first appears explaining,

> I'm called Little Buttercup,
> Dear Little Buttercup,
> Though I could never tell why,
> But still I'm called Buttercup,
> Poor Little Buttercup,
> Sweet Little Buttercup I!

Buttercup sings with somewhat contrived naïveté, but the symbolism of the buttercup's rustic innocence is probably best portrayed by the nineteenth-century English novelist, poet, and critic George Meredith. In his tragic novel *The Tale of Chloe* (1879), an old duke speaks of his wife, a milkmaid: "You should know the station I took her from was . . . it was modest. She was absolutely a buttercup of the fields."

CAMELLIA

That boasts no fragrance and conceals no thorn

Camellias are members of the tea family, the Theaceae, and are native to the eastern Asian mainland, the Himalayas, and the islands from Japan to Indonesia. Camellias are shrubs or small trees noted for their remarkable brightly colored flowers and glossy evergreen leaves. Thousands of cultivars expand the already wide selection offered by the nearly 250 species of *Camellia*. One popular species is *C. sasanqua*. The name sasanqua is the Japanese name for camellia and means wild tea flower in old Japanese. Probably the most widely known and grown camellia is *C. japonica* with at least 2000 single and double cultivars. *Camellia reticulata*, native to China, has 400 cultivars including the very large carmine-flowered 'Captain Rawes'. *Camellia sinensis* is the species whose leaves are used for tea.

Contrary to some beliefs, camellias are not named after Camille, the French courtesan of the novel *La Dame aux camélias* (1848) by Alexandre Dumas *fils*, which is also the basis for Dumas's play, Verdi's opera *La Traviata*, and a movie starring Greta Garbo. Carl Linnaeus initially named the plant *Thea sinensis*, meaning Chinese tea, in the first edition of *Species Plantarum* of 1753; however, in a later edition he named the plant in honor of Georg Joseph Kamel, a Moravian Jesuit botanist who traveled through Asia in the late 1600s. His name was Latinized as Camellus since there is no *k* in the Latin alphabet. His studies included the botany of the Philippines, particularly the flora on the island of Luzon.

According to legend, camellias came into being when an Indian prince and mystic named Dharma, the personification of truth and justice, traveled through China teaching philosophy. He became exhausted and weary from his prayer, fasting, and long travels in unfamiliar territory. One night as atonement for his weakness and inadequacy, he cut off his eyebrows and flung them to the ground. During the night the eyebrows took root and grew into the first tea plants. Dharma found that infusions or decoctions of its leaves promoted alertness, stimulated the mind, and enabled him to proceed in his teachings.

The single, wild form of the camellia drops its flower heads abruptly, suggesting sudden death. Because of this fragility, it has become a symbol of losing one's life, and traditionally camellias are not given as gifts. In the Far East they are commonly used at funerals but never at weddings. In the Victorian language of flowers, the white *Camellia japonica* denoted the sentiments of unpretending excellence and steadfastness, whereas the red-flowered form suggested loveliness. Because of its bright flowers, the camellia is sometimes called the rose of Japan.

William Roscoe, English banker, botanist, poet, and founder of the Liverpool Botanic Garden, wrote, "The chaste camellia's pure and spotless bloom/That boasts no fragrance and conceals no thorn." Mrs. C. W. Earle also appreciated the camellia. Describing her gardening efforts in diary form, she published *More Pot-Pourri from a Surrey Garden* in 1899. The entry for 8 March 1898 addresses the placement and basic needs of camellias:

> My two large old Camellias planted out last autumn, well
> under a Holly and facing north, are doing well, and one has
> three bright rosy-red blooms. It remains to be seen how they

will do next year. It is a pleasure to think Camellias do better in London gardens than almost any other evergreens,
and only want well planting in peat and leaf-mould, and
well syringing and watering in the spring. But there also they
must have the protection of other shrubs, to hang over their
tops and keep off spring frosts.

CARNATION
The fairest flowers o' the season

The genus *Dianthus* contains such garden favorites as carnations, sweet Williams, and pinks, which are members of the Caryophyllaceae. The thousands of garden cultivars of this group of 300 species have a wide range of horticultural uses, such as cut flowers, border plants, and corsages. The natural distribution of most species is through Europe and Asia. *Dianthus caryophyllus* is the clove pink, the presumed parent of all garden carnations. Also popular are the biennial sweet William, *D. barbatus*, and the cheddar pink, *D. gratianopolitanus*.

The name dianthus derives from the Latin words *di* or *dios*, in reference to the god Zeus, and *anthos*, for flower; thus the name means Zeus's flower. Stemming from the Roman name for Zeus, Jupiter or Jove, the French call the dianthus *fleur de Jupiter*. The common name carnation comes from *carne*, for flesh, in reference to the color of its flowers. The color is probably also the origin of the common name pinks, which is applied to the family of plants as a whole. The pinking shears and the expression "to pink," or snip an edging onto cloth, derive from the resemblance of the cut to the edge of the flower petals. To pink also meant to prick or cut an eyelet hole in cloth. The origin of the common name sweet William is unclear, though it is fairly clearly not a reference to William the Conquerer. Some suggest it is named after St. William, whose feast day is 21 June. Others say the name is a corruption of the French word *oeillet*, meaning eyelet, which may have been mispronounced by English speakers as "willy."

The Latin name for the family, Caryophyllaceae, is derived from the Greek *karya phyllon*, meaning walnut leaf, an allusion to the aromatic smell of the leaves of certain species. Old writings, such as those by Chaucer and Spenser, also called pinks after their strong scent, referring to them as clove gilliflowers or, because their petals were used to add a spicy flavor to wine, sops-in-wine. The name gilliflower comprehensively referred to several different flowers: carnations, wallflowers, and pinks. It may come from *giroflée*, French for clove, which the flower's scent resembles. Others suggest that the name gilliflowers, or sometimes gilly flowers, is an erosion of "July flowers," because that is when they bloom. Adding further to linguistic confusion, modern French applies the common name *giroflée* to the wallflower (formerly *Cheiranthus cheiri* but now *Erysimum cheiri*), a member of the mustard family.

Although growing in European countries for centuries, pinks and their botanical allies were considered the flowers of the Renaissance, because not until then were they "discovered," grown in profusion, and widely esteemed. They represent the glory, energy, and vitality of that era. For example, Botticelli and others painted pinks, and people sat for portraits with pinks in their hands. John Parkinson wrote a long chapter on dianthus in his *Paradisi in Sole, Paradisus Terrestris* (1629), because it was one of "the chiefest flowers of account in all our English gardens."

The earliest known reference to the clove gilliflower in English literature is in Chaucer's *Tale of Sir Thopas* (1394), in which the title character "priketh thrugh a fair forest" in Flanders and observes,

> There springen herbes grete and smale,
> The lycorys and cetewall,

And many a clove-gilofre
And notemuge to putte in ale,
Whether it be moyste or stale.

References to carnations and other pinks began appearing in literature more frequently in the sixteenth century at the peak of the Renaissance. In sonnet 63 of *Amoretti* (1595), Edmund Spenser used the
beauty of flowers to describe a damsel:

Her lips did smell lyke unto Gillyflowers,
Her ruddy cheekes lyke unto Roses red,
Her snowy browes lyke budded Bellamoures,
Her lovely eyes lyke Pincks but newly spred.

In *The Muses Elyzium* (1630), Michael Drayton, an English poet and
playwright and contemporary of Shakespeare, wove a garland of the
July flower:

My chiefest flower this make I:

. . . .

The curious choice clove July flower
Whose kinds height the carnation,
For sweetness of most sovereign power
Shall help my wreath to fashion.

In Shakespeare's *The Winter's Tale* (1610), Perdita, left by Antigonus as a
babe on a lake shore in Bohemia, is taken in by an old shepherd and
raised as his daughter. In act 4, while serving as "hostess-ship o' the
day" to Old Shepherd and others, Perdita says to him:

Sir, the year growing ancient,
Not yet on summer's death, nor on the birth
Of trembling winter, the fairest flowers o' the season
Are our carnations and streaked gillyvors,
Which some call nature's bastards: of that kind

Our rustic garden's barren; and I care not
To get the slips of them.

Milton's elegant paean "Epitaph on the Marchioness of Winchester" (1631) describes the wife of the Earl of Winchester who became ill and died at the age of twenty-three, during her second pregnancy. Hanging over her deathbed are faded flowers from her wedding day. Carnations adorn her funeral bier:

The pride of her carnation train,
Plucked up by some unheedy swain,
Who only thought to crop the flower
New shot up from vernal shower.

English essayist and metaphysical poet Abraham Cowley in "Sweet William" described the bloom as Jove's flower with a beard:

Sweet-William small, his form and aspect bright;
Like that sweet flower that yields great Jove delight;
Had he majestic bulk he'd now be styled
Jove's flower; and, if my skill is not beguiled,
He was Jove's flower when Jove was but a child.
Take him with many flowers in one conferred,
He's worthy Jove, ev'n now he has a beard.

In Shakespeare's *Romeo and Juliet* (1597), the term "pink of courtesy" makes an early appearance in a punned allusion to the flower:

ROMEO: A most courteous exposition.
MERCUTIO: Nay, I am the very pink of courtesy.
ROMEO: Pink for flower.
MERCUTIO: Right.
ROMEO: Why, then, is my pump well flowered.
MERCUTIO: Well said: follow me this jest now till thou hast

worn out thy pump, that when the single sole of it is
worn, the jest may remain after the wearing sole singular.

An old folk song alludes to both pinks, the flower, and the pink of
perfection. Robert Burns recorded it in "The Posie" (1794):

O, luve will venture in whaur it daur na weel be seen!
O, luve will venture in where wisdom once has been;
But I will doun yon river roue among the woods sae green,
An' a' t' pu' a posie t' my ain dear May!
The primrose I will pu', the firstling of the year,
An' I will pu' the pink, the emblem o' my dear,
For she's the pink o' womankind, an' blooms without a peer;
An' a' t' be a posie t' my ain dear May.

Carnations are said to have played a role in the near escape of Marie
Antoinette during the French Revolution. In September 1793, a loyal
supporter communicated the escape plan by hiding a message in a carnation and dropping the flower at her feet. Not having a pen, Marie
Antoinette pricked her reply onto a scrap of paper, but the jailers discovered the plot and the escape failed.

In some parts of the United States carnations are worn on Mother's Day, when a red flower shows that one's mother is living, and a
white one shows that she has passed away. The language of flowers
has a wide range of meanings for the many forms of carnations. For
example, the striped carnation means "Alas, my poor love"; a single
pink suggests pure love; and the sweet William represents boldness.
The immense popularity of the pink family members has waned a bit
since the sixteenth and seventeenth centuries, but from this twentieth-
century vantage point they are a clear symbol of that era's vitality.

CLEMATIS

Most beauteous with its feathery plume

About 230 herbaceous to woody species of *Clematis* are found in the Ranunculaceae. These perennials are distributed in the temperate regions worldwide and in the mountains of equatorial Africa. Most are climbers; some are evergreen, for example, the well-known garden species *C. armandii* from southwest China. *Clematis virginiana* is a deciduous, semi-woody plant from eastern North America.

The name *Clematis* is from the classical Greek word *clema*, meaning tendril or vine branch, which was applied to various climbing plants.  One of its common names, virgin's bower, does not allude to Elizabeth I, the Virgin Queen; the name was used for the plant before her time. In France, an old name for clematis was *de la Sainte Vierge*, suggesting a relationship to the Virgin Mary. *Clematis vitalba*, the native English species common to hedgerows, often appears in literature as traveler's joy. According to Gerard, "It was used as decking and adorning waies and hedges where people traveil." *Clematis vitalba*, whose specific name means white vine, is also called old man's beard from the effect of the feathery gray styles left on the fruit after the flower has faded. Another common name, woodbine, is variously applied both to clematis and to *Lonicera*, the honeysuckle. Since both are climbers, when woodbine is mentioned in literature it is sometimes difficult to know which plant was intended.

Several different forms of the clematis had individual meanings in the language of flowers. In general, the plant signified mental beauty. The evergreen clematis meant poverty. The traveler's joy, *Clematis vitalba*, conveyed safety. Another meaning assigned generally to the clematis was artifice. Beggars and gypsies induced this significance by their use of the plant's acrid leaves to inflict ulcers and sores on their bodies to invoke charity and compassion.

John Parkinson introduced the world of climbing plants in his chapter "Clematis: Clamberers, or Creepers" from *Paradisi in Sole, Paradisus Terrestris* (1629):

> Having shewed you all my store of herbes bearing fine
> flowers, let mee now bring to your consideration the rest
> of those plants, be they shrubs or trees, that are cherished
> in our garden, for the beauty of their flowers chiefly, or for
> some other beautiful respect; and first, I will begin with
> such as creepe on the ground, without climing, and then
> such as clime up by poles, or other things; that are set or
> grow neere them, fit to make Bowers and Arbours; or else
> are like them in forme, in name, or some other such quality
> or propertie.

He went on to mistake periwinkle for a creeping clematis and described *Clematis virginiana* (probably a *Passiflora*) as "the Virginia Climer . . . such as hath been brought to us from Virginia. . . . [and] the bud of the flower, before it doe open, is very like unto the head or seede versell of the ordinary single *Nigella*."

The clematis in literature trails over rustic arbors and creates secluded bowers. A poem by John Clare called "Proposals for Building a Cottage" (ca. 1820) mentions woodbine, meaning the honeysuckle, and old man's beard, *Clematis vitalba*:

A little garden, not too fine,
Inclose with painted pales;
And woodbines, round the cot to twine,
Pin to the wall with nails.

Let hazels grow, and spindling sedge,
Bent bowering overhead;
Dig old man's beard from woodland hedge,
To twine a summer shade.

In *The Lady of the Lake* (1810), Sir Walter Scott drew this verdant scene:

Due westward, fronting to the green,
A rural portico was seen,
Aloft on native pillars borne,
Of mountain fir, with bark unshorn,
Where Ellen's hand had taught to twine
The ivy and Idaean vine,
The clematis, the favour'd flower
Which boasts the name of Virgin-bower,
And every hardy plant could bear
Loch Katrine's keen and searching air.

Other English writers have described the joy of clematis. In a journal entry of 23 November 1788, Gilbert White wrote, "The downy seeds of travellers joy fill the air & driving before a gale appear like insects on the wing." Bishop Richard Mant penned, "The Traveller's Joy,/Most beauteous when its flowers assume/Their autumn form of feathery plume." Rosamund Marriott Watson, using the pseudonym Graham R. Tomson, described the sweet traveler's joy in "Vespertilia":

Over the hills and far away

The road is long on a summer's day;
Dust glares white in the noontide heat,
But the Traveller's Joy grows strong and sweet,
Down the hollow and up the slope
It binds the hedge with a silken rope.
O the sun that shines and the dust that flies,
And the fresh green leaves for tired eyes—
Green leaves, and the summer's hope.

In *Endymion* (1817), John Keats described both virgin's bower and woodbine in the story of a "brain-sick shepherd-prince" of Mount Latmos who falls in love with Cynthia, the moon, and travels the earth to find her. On his travels, Endymion finds in a myrtle-walled bower the sleeping youth Adonis "of fondest beauty":

Together intertwin'd and trammell'd fresh:
The vine of glossy sprout; the ivy mesh,
Shading its Ethiop berries; and woodbine,
Of velvet leaves and bugle-blooms divine;
Convolvulus in streaked vases flush;
The creeper, mellowing for an autumn blush;
And virgin's bower, trailing airily.

COCKSCOMB

From a golden cockerel

The cockscomb comprises about fifty species in the Amaranthaceae, the most common of which is *Celosia argentea*. The name *Celosia* is Greek for burning, an allusion to the brillantly flame-colored flower heads of the genus. The common name likewise refers to the flower heads, which on some species resemble the fleshy crest or caruncle of domestic fowl.

How the cockscomb got its name is told in a Korean legend about two village neighbors, Kim and Choi. Among the fowl that Kim owned was an especially prized golden cockerel. One day Choi asked Kim to play checkers. Choi won and suggested that their cockerels fight; having lost the game of checkers, Kim was obliged to agree to the cockfight. Choi attached hidden knives to the legs of his cockerel, and Kim's golden one was killed. Kim and his wife sadly buried their prized bird in their yard. The next morning as Kim was sweeping the yard, he found a cockscomb flower growing where the golden cockerel was buried.

COLCHICUM

Warms the cold bosom of the hoary year

Forty-five species of *Colchicum*, members of the Liliaceae (or the smaller, more recently defined Colchicaceae), are distributed in the Northern Hemisphere in Europe, northern Africa, and western Asia (some taxonomists include species of *Bulbocodium* and *Merendera* as *Colchicum* species, thereby raising the total to about sixty-five). They are herbaceous plants that grow from corms and produce stalkless flowers, usually before the leaves emerge. Colchicums are commonly called autumn crocus or meadow saffron because of their visual similarity to the crocus. But they are botanically easily distinguishable—colchicums have six stamens whereas crocuses have only three, putting the two plants in different families (see the chapter "Crocus"). Colchicums are variously called naked ladies, naked boys, or upstart because the flowers of most species emerge without the benefit of foliage. *Colchicum autumnale* is the autumn-blooming colchicum, which has goblet-shaped flowers of pale purple to white. *Colchicum speciosum*, which normally is lilac to pink, also has a dull white-colored form, 'Album'. Two spectacular cultivars are the double-petaled *Colchicum* 'Water Lily' and the richly hued 'Violet Queen'.

Colchicums are the source of the alkaloid colchicine, which is used in plant genetics and breeding to produce polyploidy, a doubling of the chromosomes, during cell division. The chemical formerly was used as a medicine to treat gout. Some varieties of plants with large flowers or doubled petals have been produced by the use of colchicine. Old French names for the colchicum are *tue-chien* and *mort-au-chien*, meaning dog killer, because of the plant's poisonous attributes. It was also called, according to John Parkinson, "the sonne before the father" (*filis ante patrem* in Latin) because flowers appear before the leaves. But as Parkinson said, "This is without due consideration, for the root of this (as most other bulbous plants) after the stalke of leaves and seede are dry, may be transplanted, and then it beginneth to spring and give flowers before leaves."

The name colchicum is from the city of Colchis, near Georgia on the Black Sea, where the plant historically grew and, according to myth, where the Argonauts found the Golden Fleece. Horace in his writings about the Euxine, the old name for the region of the Black Sea, noted that in the land of Colchis the colchicums grew in great abundance: "Or tempered every baleful juice/Which poisonous Colchican glebes produce." Myths tell that Medea prized the colchicum and used it in her sorcery. In one account, Medea fell in love with Jason in Colchis and concocted a brew to restore his youth. As drops of the liquor fell on the ground, colchicums sprang up. Erasmus Darwin in *The Loves of the Plants* (1789) described Medea's potion:

> So when Medea to exulting Greece
> From plunder'd Colchis bore the golden fleece,
> On the loud shore a magic pile she raised,
> The cauldron bubbled, and the faggots blazed;
> Pleased, on the boiling wave old Aeson swims,
> And feels new vigour stretch his swelling limbs.

In the same work, Darwin narrated the autumn flowering of the colchicum:

> Then bright from earth, amid the troubled sky,
> Ascends fair Colchica with radiant eye,
> Warms the cold bosom of the hoary year
> And lights with beauty's blaze the dusky sphere.

William Cobbett, an English farmer, member of parliament, and radical (he was imprisoned for two years for his writings on flogging in the army), penned his thoughts and reflections in a plain, broad style. He was singularly excited about the beauty of the colchicum he saw in Gloucestershire near Bollitree in September 1826. He wrote of it in *Rural Rides* (1830), a collection of essays that rails against taxes and landlords and reveals his uncomplimentary view of London through his discussion of the London wen, a cyst formed by the obstruction of a sebaceous gland. Despite the essay's critical viewpoint, it includes lovely images of the rural landscape:

> As I came along, I saw one of the prettiest sights in the flower way that I ever saw in my life. It was a little orchard; the grass in it had just taken a start, and was beautifully fresh; and very thickly growing amongst the grass, was the purple-flowered colchicum in full bloom. They say that the leaves of this plant, which come out in the spring and die away in the summer, are poisonous to cattle if they eat much of them in the spring. The flower, if standing by itself, would be no great beauty; but, contrasted thus with the fresh grass, which was a little shorter than itself, it was very beautiful.

In keeping with the late and short blooming of the flower, colchicums said "My best days are over" in the language of flowers.

COLUMBINE

The war and peace flower

The columbine belongs to the genus *Aquilegia* within the Ranunculaceae. The seventy to eighty species are native to the Northern Hemisphere. The common columbine or granny's bonnet of Europe is *A. vulgaris*. *Aquilegia canadensis* with red petals and yellow spurs and the purple-blue *A. jonesii* are cultivated North American species. Because *Aquilegia* species hybridize freely, numerous seed strains are available.

The Latin and common names of this plant provide an oxymoronic lesson in ornithology. The name columbine owes its origin to *columba*, for dove, since some fancied a resemblance between a stand of the inverted flowers and a flock of doves. The Latin word is also the root of the words columbary, or dove-cote, and columbarium, a vault of funerary urns placed in niches reminiscent of a dovecote. Linnaeus provided the genus with its Latin name, *Aquilegia*, which refers to *aquila*, or eagle, because the spurs on the flower look like talons of the lordly bird of prey. Thus, although its common name is a symbol for peace, its botanical name is a symbol for war. Others suggest that the Latin names come from the root word aqua, meaning water, because the shape of the flower suggests a funnel for bearing water. An old name for columbine was *herba leonis* in the belief that it was the favorite herb of the lion.

In the language of flowers, several sentiments were represented by the columbine. The wild columbine meant folly, the purple columbine suggested resolution, and the red columbine suggested anxiety and trembling. In the church and particularly in religious art, a stalk of columbine represented the Holy Ghost, and a stalk with seven flowers suggested the seven gifts of the Holy Ghost (Isa. 11:2): wisdom, understanding, counsel, fortitude, knowledge, piety, and fear of the Lord.

John Parkinson wrote of the many "sorts of Colombines, as well differing in forme as colour of the flowers, and of them both single and double carefully noursed up in our Gardens, for the delight both of their forme and colours." Chaucer wrote of columbines as early as 1386 in *The Marchantess Tale*: "The turtle's voys is herd, my dowue sweete / The winter is goon, with all his reynes wete;/Com forth now with thyne eyen columbyn! . . . The gardin is enclosed al aboute."

Columbine or Columbina was the sweetheart of Arlecchino, both of whom were stock characters in Italian commedia dell'arte of the sixteenth through eighteenth centuries. She was usually a maidservant or sometimes a girlfriend whose name meant dove-like or coquettish. Writing in 1595 in *Amoretti* (sonnet 64), Edmund Spenser compared a woman's beauty to various plants, including the columbine:

> Her goodly bosom lyke a strawberry bed,
> Her neck lyke to a bunch of Cullambynes.
> Her brest lyke lillyes, ere theyr leaues be shed,
> Her nipples lyke yong blossomd [J]essemynes.

But at one time the columbine was regarded as a symbol of infidelity and may be what William Browne had in mind in the 1800s when he wrote in *Britannia's Pastorals*, "The columbine by lonely wanderer taken,/ Is there ascribed to such as are forsaken."

In Shakespeare's *Hamlet* (1601), Ophelia, in madness over Hamlet's rejection of her, cries out to her brother Laertes, "There's fennel for you and Columbines"—the fennel suggesting flattery and the columbine ingratitude. English playwright George Chapman also referred to the columbine as a symbol of ingratitude: "What's that—a columbine? No, that thankless flower grows not in my garden."

At least one observer found the similarity between the shape of the flower and the shape of a court jester's cap, which may point to the plant's suggestion of folly in the language of flowers. L. A. Twamley, the Australian Louisa Anne Twamley Meredith who wrote on Tasmanian flora in the 1800s, described the columbine as a flower of folly:

> Why, when so many fairer shine,
> Why choose the homely columbine?
> 'Tis Folly's flower, that homely one,
> That universal guest,
> Makes every garden but a type
> Of every human breast;
> For though ye tend both mind and bower,
> There's still a nook for Folly's flower.

Ralph Waldo Emerson gave an uplifting nineteenth-century tribute to the columbine in "Musketaquid":

> I am a willow of the wilderness,
> Loving the wind that bent me. All my hurts
> My garden spade can heal. A woodland walk,
> A quest of river grapes, a mocking thrush,
> A wild rose or a rock-loving columbine
> Salve my wounds.

And an unknown author wrote,

> Skirting the rocks at the forest edge
> With a running flame from ledge to ledge,
> Or swaying deeper in shadowy glooms,

A smouldering fire in her dusky blooms;
Bronzed and molded by wind and sun,
Maddening, gladdening every one
With gipsy beauty full and fine,
A health to the crimson columbine.

John Clare described the flower as "The Columbine, stone-blue, or deep night-brown,/Their honey-comb-like flowers hanging down." And later, probably in the Victorian era, an English poet described the lanky growth habit of the columbine:

In pink or purple hues arrayed, ofttimes indeed in white,
We see, within the woodland glade, the Columbine delight;
Some three feet high, with stem erect, the plant unaided grows,
And at the summit, now deflect, the strange-formed flower
   blows.

The columbine also appears in Iroquois legend. The five-petaled North American native columbine, *Aquilegia canadensis*, came about from five Indian chiefs who were changed into the flower by the Great Spirit. The chiefs had neglected their lands and their people to search for a sky maiden whom they had fallen in love with in a dream. The colorful flower represents the five chiefs huddled together in a ring wearing yellow moccasins and doeskin shirts dyed red and yellow.

CORNFLOWER

Like a cornflower in a field of grain

The cornflower and its relatives make up the 450 or so species of the genus *Centaurea* that are distributed primarily in the Northern Hemisphere. The common cornflower or blue-bottle is *C. cyanus*; it is also called bachelor's buttons in some regions. Its bluish purple daisy-like heads with distinct rays identifies it as a member of the Compositae (Asteraceae). The perennial cornflower is *C. montana*, and the lesser knapweed is *C. nigra*.

Linnaeus took the name *Centaurea* from the Greek word *centaur*, the half-man, half-horse of mythology. Chiron, the centaur, used a cornflower to heal his foot. Hercules had wounded him with the shot of one of his arrows poisoned with the blood of Hydra. Some writers suggest the plant used was *Centaurium*, or centaury plant, a member of the Gentianaceae and botanically unrelated to cornflower. Chiron also is fabled for having taught humans about the healing properties of herbs.

The specific epithet *cyanus* means dark blue. In mythology, Cyanus was a handsome if odd youth who garlanded himself in fanciful cornflowers and blue clothing. One day Flora, the goddess whom he adored, found him dead in a cornfield and transformed his body into the cornflower in honor of his veneration of her. In Russia, the cornflower is called the *basilek*, or flower of Basil, based on the tale of

Rusalka, a nymph who lured the young Basilek, or Basil, into a corn-field and transformed him into a cornflower.

The common name cornflower is suggested from the plant's frequent occurrence in fields of cereal grain; as John Parkinson wrote, it can be found "furnishing or pestering the Corne-fields." (Throughout Europe, the term corn generally refers to wheat, barley, oats, and rye; in North America, corn means Indian corn, or *Zea mays*.) Francis Bacon also noted the cornflower in *Sylva Sylvarum* (1626), a collection of natural history observations and phenomena: "There be certain Corn-flowers which come seldome or never in other places but onely amongst Corn: As the blew Bottle, a kind of Yellow Mary-Gold, Wilde Poppy and Fumitory . . . . It should seem to be the corn that qualifieth the earth, and prepareth it for their growth." *Centaurea* has a surfeit of other common names including knapweed, star thistle, and blue-bottle. In some regions of England, it was once known as blunt-sickle and hurt-sickle, because the tough flower stems blunted the reaper's hand-held scythe when grain was being cut.

John Parkinson described the uses of "Cyanus, corneflower, or blew Bottles":

> There is no use in Physicke in Galen and Dioscorides time, in that (as it is thought) they have made not mention of them; we in these dayes doe chiefly use the first kindes (as also the greater sort) as a cooling cordial, and commended by some to be a remedy not only against the plague and pestilential diseases, but against the poison of scorpions and spiders.

In the language of flowers, the cornflower was a symbol of delicacy or sensitivity, reminiscent of the devotion of Cyanus to Flora. A single bachelor's button flower suggested celibacy, but why the cornflower began to be called bachelor's button is not clear. Some suggest that the flower resembles buttons worn on men's clothing in the sixteenth

century. Too, there are traditions in which wearing the flower indicates
that one is unmarried. Picking the flower while dew is still on it and
wearing it for a full twenty-four hours will predict one's success in
courtship, depending on the plant's condition. This practice may be
the origin of the expression "true blue."

In *The Merry Wives of Windsor*, which Shakespeare wrote in 1597, the
host of the Garter Inn teases Fenton, a gentleman, about his love life
because he is wearing a button of flowers:

> What say you to young Master Fenton? he capers, he
> dances, he has eyes of youth, he writes verses, he speaks
> holiday, he smells April and May: he will carry 't, he will
> carry 't; 'tis in his buttons; he will carry 't.

In his long poem "The Village Minstrel" (1821), John Clare compared the minstrel to the random, conspicuous, unwanted appearances of the cornflower:

> He has friends, compar'd to foes though few,
> And like a cornflower in a field of grain
> 'Mong many a foe his wild weeds ope to view,
> And malice mocks him with a rude disdain.

CRAPE MYRTLE

The one-hundred-day red flower

The crape myrtle is *Lagerstroemia*, a member of the Lythraceae. The fifty or so species are trees and shrubs native to tropical Asia and Australia. The genus is named for Magnus von Lagerström, a Swedish merchant and friend of Carl Linnaeus. The name crape is the Americanization of the word crepe, which is from the Latin *crispus*, for curly, because of the ruffled margins of the crape myrtle's flower petals. The word crape is also related orthographically to crepe paper and to crepe, the thin worsted cloth formerly worn by the clergy, two substances reminiscent of the petals of the crape myrtle. The name myrtle alludes to the myrtle shrub, *Myrtus*, which the crape myrtle superficially resembles when not in bloom.

In Korea, the crape myrtle is called one-hundred-day red flower, a name it gained through legend. Once, a three-headed sea dragon demanded of a coastal village that each year a maiden be given in sacrifice as his bride. The house of Kim had a beautiful daughter, and one year the townspeople decided that she would be sacrificed to the dragon. She was dressed in a wedding gown and taken to the shore. As the dragon came to take her into the sea, a prince sailed in and cut off one of the sea dragon's heads, thereby saving the maiden. The prince's father arranged for the prince and the maiden to marry, but just as the wedding day arrived, the king's treasures disappeared. The king sent word to his son that he was unhappy and could not go through with the ceremony. The prince promised to find the treasures within 100 days and return to marry his betrothed. As he bade farewell to her, he said he would fly a white flag from his ship if the trip were successful and a red one if not. The maiden waited 100 days, and finally the prince's ship was spotted. When the maiden looked for the signal flag,

she spied a red flag flying on the ship and died from the grief of knowing that she would not be wed to the prince. But when the prince arrived, those in the village could see that the red flag was truly a white
flag stained with the blood of the dragon that the prince had slain to
regain the king's treasure. The victorious prince arrived just in time to
attend his maiden's funeral. From her grave sprang the crape myrtle,
which blooms 100 days each summer.

CROCUS

Thou venturous flower

The eighty or more species of *Crocus* belong to the Iridaceae, the same family as the iris. Their distribution is from south-central Europe, through the Middle East, to central Asia. None are native to North America. Though they are thought of as spring flowers, several species are autumn-blooming crocus, such as *C. speciosus* and *C. goulimyi*. The colchicum, however, which sometimes bears the common name autumn crocus, is botanically unrelated, being a member of the lily family, or the smaller, more recently defined colchicum family, Colchicaceae (see the chapter "Colchicum"). Early naturalists recognized only two forms of crocus, *C. sativus*, the saffron crocus, and *C. vernus*, the spring-flowering crocus; all the rest were thought to be varieties of them. The large, opulent purple and white crocuses of our spring gardens are selections of *C. vernus*.

The name crocus is derived from the old Greek word *krokos*, for saffron, the flavoring and confectionery coloring; its source is the orange-colored stigmata of the saffron crocus. The word is ancient and appears in various forms in other Middle Eastern languages, with variable meaning. One meaning relates the word to thread or fiber, since the saffron was used for dyeing textiles. Some etymologists have suggested that the words crocus and crocodile have a similar origin. Thomas Fuller explained their connection in *The Worthies of England* (1662), which assiduously catalogs the important people, industries, proverbs, churches, and other items "painful and pious" in the counties of England. He said the following of the saffron crocus: "In a word, the sovereign power of saffron is plainly proved by the antipathy of the crocodiles thereunto: for the crocodile's tears are never true, save when he is forced where saffron groweth (whence he hath his name χροκο-δειλοσ, or the saffron-fearer) knowing himself to be all poison, and it all antidote." Numerous forms of crocus are familiar in our spring and autumn gardens. The saffron, whose name comes from the Spanish *azafaran*, which itself comes from the Arabic *za'faran*, is a poor garden performer because it has sparse flowers and is disease prone.

Crusaders returning from the Holy Land introduced the saffron crocus to the banquet table of Henry I of England, but the plant also appears to have entered Britain via the Romans. At least one town, Saffron Walden in Essex, took its name from the area's former chief commodity, production of which was at its height in the 1700s, offering good quality saffron. Thomas Fuller's treatment of saffron was prompted by the industry's English presence. By the time of Henry VIII, bed sheets were dyed with saffron to make them antiseptic; but he ultimately banned its use as a dye on linen because of the expense.

In the language of flowers, the crocus suggested mirth, the gladness of youth, and the pleasure of hope. The Doctrine of Signatures suggested using saffron to cure jaundice and urinary problems.

For centuries the saffron has served as a medicine (with little proven benefit, although it is rich in vitamin B_2) and as a commercial commodity as a dye, foodstuff, and perfume. On the island of Crete the plant's image appears in Minoan designs on faience and other earthenware, on costumes, and on decorative motifs in temples, where crocus appears to have had religious significance particularly to the Minoan goddess Britomartis. At the palace of Minos on Crete, a fresco referred to as *The Blue Boy Picking Crocus* dates from about 1900 BCE. Influenced by overseas traders, a flourishing industry of the luxurious, bright orange-red stigmata developed on the shores of Crete. Because about 4400 stigmata are needed to make an ounce of saffron, the hand processing may have fueled a Minoan economy of great proportions.

Legend suggests that Crocus was a beautiful youth from the plains who was passionate for the hill-dwelling shepherdess Smilax. Because they could not marry, they both prayed to the goddess Flora, who helped them find love by transforming them into flowers. As plants, the tendrils of the smilax bound garlands of crocus. The Greeks used this combination in bouquets at marriage festivals. René Rapin, perhaps borrowing from Ovid's earlier version of the legend, told of the horticultural match: "Crocus and Smilax, once a loving pair,/But now transformed, delightful blossoms bear." In another legend, the gods came down from Mt. Olympus to a meadow to relax and play games one day. The speedy and winged Mercury tossed a disc that went off course, killing Crocus, the infant son of Europa. The crocus flower sprang from the blood of the babe.

The crocus is dedicated to St. Valentine, who lived under Roman rule in the third century. Valentine was a physician who practiced medicine with various herbs and powders. One day a jailer presented his young blind daughter to Valentine and asked if he could cure her. Over a period of several weeks, as Valentine administered a waxy ointment to her eyes, he befriended the child. The two often took walks together in the nearby fields where the child would pick bouquets of crocuses to give to her father. During an uprising in the streets of Rome, Christians were persecuted and Valentine was put in jail. As he was being prepared for execution on 14 February 270, he wrote a note to give to the blind young girl. When she opened it, out fell a pressed crocus, which she was immediately able to see. The note read, "From your Valentine." Valentine was accorded sainthood on 14 February 496 by Pope Gelasius I. Each year on this date the crocus is said to bloom at dawn. Saint Valentine's feast day is related to the much older Roman Feast of the Lupercalia, a fertility festival that was celebrated a day later on 15 February.

The crocus was written of as early as ca. 700 BCE, when Homer wrote that the love couch of Jove and Juno was adorned with nuptial crocus flowers. The saffron crocus is believed to be the *karkum* men-

tioned in the original Hebrew of the biblical Song of Solomon (4:13):
"The plants are an orchard of pomegranates, spikenard and saffron;
calamus and cinnamon, with all trees of frankincense; myrrh and
aloes, with all the chief spices." Later, in 1610, Shakespeare included
crocus in the list of items that Clown, the son of Old Shepherd, needs
to buy for the sheep-shearing feast in *The Winter's Tale*: "I must have
saffron to colour the warden pies; mace; dates?—none, that's out of
my note; nutmegs, seven; a race or two of ginger, but that I may beg;
four pounds of prunes, and as many of raisins o' the sun."

   In 1665 René Rapin in *De Hortorum Cultura* placed crocus within the
progression of the garden's vernal wakening:

>   And next the Crocus, from whose slender Bloom,
>   The Flow'rs which blow a doubtful Paint assume
>   And through the rural Plains new Fragrance spread
>   On a weak Stem scarce holds her Bending head.

Many writers have marveled at the brave little crocus that bursts forth
in the spring, often with snow still on the ground. Late in the seventeenth century Matthew Prior wrote "To the Crocus":

>   Dainty young thing
>   Of life! thou venturous flower,
>   Who growst through the hard cold bower
>   Of wintry spring.
>   Thou various hued,
>   Soft, voiceless bell, whose spire
>   Rocks in the grassy leaves, like wire
>   In solitude.

   Regarding the distinctive purple crocus, in the nineteenth century
Mary Botham Howitt wrote in "The Wild Spring Crocus" from *Sketches
of Natural History* (1847):

>   Like lilac-flame its colour glows,

Tender and yet so clearly bright,
That all for miles and miles about
The splendid meadow shineth out;
And far-off village children shout
To see the welcome sight.

And Henry Wadsworth Longfellow added yellow crocus to Christ's crown in "The Golden Legend":

Hail to the King of Bethlehem!
Who weareth in His diadem
A yellow crocus for the gem
Of His authority.

Both the spring- and autumn-flowering crocus are mentioned in these anonymous lines:

Say what impels, amid surrounding snow
Congeal'd, the crocus' flaming bud to grow?
Say, what retards, amidst the summer's blaze,
Th' autumnal bulb, 'til pale, declining days?

And the crocus appears in *Wuthering Heights* (1847) by Emily Brontë. Nelly Dean, Heathcliff's housekeeper, says:

That Friday made the last of our fine days, for a month. In the evening the weather broke; the wind shifted from south to north-east, and brought rain first, and then sleet and snow. On the morrow one could hardly imagine that there had been three weeks of summer: the primroses and crocuses were hidden under windy drifts; the larks were silent, the young leaves of the early trees smitten and blackened. And dreary, and chill, and dismal, that morrow did creep over!

The crocus is generally considered the first flower of the spring season. We treasure it for its bravery in boldly peering out of the earth ahead of other flowers. Its 4000 years of human association confirms its place in horticultural history.

CYCLAMEN

With robes of silver born

The genus *Cyclamen* consists of twenty tuberous species in the Primulaceae. They are native to western Asia, Europe, and northern Africa. The common showy greenhouse cyclamen is derived from *C. persicum*, which is native to the eastern Mediterranean. The hardier species such as *C. hederifolium*, *C. coum*, and *C. purpurascens* have attractive leaf patterns and have become popular in the United States in the last decade. They provide small but showy flowers, particularly in the autumn and winter.

The name cyclamen is derived from the Greek *kyklos* or *cyclos*, for circular, referring to the coiling of the flower stems after pollination, the shape of the leaves, or perhaps the flattened, round, corm-like tubers. The name cyclamen and variations of it have been used since at least the time of Theophrastus in the third century BCE. The common names sowbread or swine bread, in English and other languages (*pain de porceau* in French, for example), come from the belief that the plant afforded food for wild swine. The specific name of the well-known hardy *Cyclamen hederifolium* means ivy-leaved, from the resemblance of its foliage to the genus *Hedera* (ivy). Another but perhaps less well-known common name for *C. hederifolium* is bleeding nun; like many other flowers it was consecrated to the Virgin Mary. The species name of the winter-blooming *C. coum* is generally believed to mean "from the Island of Kos" in the Aegean Sea, but the species has not been found there.

William Turner appears to have been the first to use the name sowbread for cyclamen in *A New Herball*, compiled from 1551 to 1568. Although he himself did not see the plant growing in England, he thought it might be found there and that if it were it ought to have a name: "Least it should be nameles, if ether it should be brought into England, or be found in anye place in England, I name it Sowesbreade." John Gerard, who grew both *Cyclamen hederifolium* and *C. coum*, wrote in his *Herbal or General History of Plants* (1633 revision by Thomas Johnson), "The common kinde of Sowbread . . . hath many greene and round leaues." In his 1830 *Introduction to the Natural System of Botany*, John Lindley explained that although sowbread is "famous for its acridity, yet this is the principal food for the wild boars of Sicily." Additional insight on the plant's name is given by Philip Miller in *The Gardener's Dictionary* (1731): "It is called Sowbread because the Root is round like a Loaf, and the Sows eat it."

Cyclamens, primarily *Cyclamen hederifolium*, were used for treatment of a number of complaints and were known by the apothecary's name of *panis porcinus*, Latin for sow or swine bread. They were used to stop baldness, to counteract certain poisons, and to aid in childbirth (sow bread was considered a potent assistant by midwives and was believed to help women while in confinement). Cyclamens were also an "amorous medicine" used to bestir voluptuous desires. Owen Wood, in a book with the cumbersome title of *An Alphabetical Book of Physicall Secrets, for All Those Diseases That Are Most Predominant and Dangerous in the Body of Man* (1639), wrote, "Sowbread root . . . with honied water, purgeth grosse phlegme and filthy humours." Abraham Cowley wrote in *Sex Libri Plantarum* (*Lives of the Plants*), translated by Aphra Behn in 1687, "The Sow-Bread does afford rich Food for Swine, Physick for Man, and Garlands for the Shrine."

In the language of flowers cyclamens signify diffidence and distrust. Thus, there may be hidden meaning in an old ballad, because the cyclamen was intended to protect the bed chamber while its occupant slept:

> St. John's Wort and fresh cyclamen—
> She in her chamber kept,
> From the power of evil angels
> To guard him while he slept.

English poet and literary critic Walter Savage Landor wrote at the beginning of the nineteenth century in a style often imitative of Greek and Latin works, such as in this ode "To the Cyclamen":

> Thou cyclamen of crumpled horn
> Toss not thy head aside;
> Repose it where the Loves were born,
> In that warm dell abide.
> Whatever flowers, on mountain, field,
> Or garden, may arise,
> Thine only that pure odor yield
> Which never can suffice.
> Emblem of her I've loved so long,
> Go, carry her this little song.

Katherine Bradley and Edith Cooper, aunt and niece who wrote together under the pen name Michael Field, described the impact of the flowers in "Cyclamens":

> They are terribly white:
> There is snow on the ground,
> And a moon on the snow at night;
> The sky is cut by the winter light;
> Yet I, who have all these things in ken,

> Am struck to the heart by the chiseled white
> Of this handful of cyclamen.

Ralph Waldo Emerson wrote *English Traits—First Visit to England* (1856), an insightful assessment of the character of English people. The work describes his 1833 trip to France, Italy, and England; in Florence he met Horatio Greenough, the American sculptor, who introduced him to Mr. Landor,

> who lived at San Domenica di Fiesole. On the 15th May
> I dined with Mr. Landor. I found him noble and courteous,
> living in a cloud of pictures at his Villa Gherardesca, a fine
> house commanding a beautiful landscape. I had inferred
> from his books, or magnified from some anecdotes, an
> impression of Achillean wrath,—an untamable petulance.
> I do not know whether the imputation were just or not,
> but certainly on this May day his courtesy veiled that
> haughty mind, and he was the most patient and gentle of
> hosts. He praised the beautiful cyclamen which grows all
> about Florence.

John Ruskin, writing in *Proserpina* (completed 1886), his studies of wayside flowers, called the cyclamen "the most capricious of all beautiful wild flowers." He went on to say that this pretty flower is "insultingly said by all nations to be good to feed pigs," and he called it the "pig turnip." "Yet all the while I have never heard of [anyone] growing fields of cyclamen for their pigs, nor of pigs routing in the fields for roots of cyclamen."

René Rapin eloquently explained the origin and habit of the cyclamen:

> Cyclamens, which we now with pleasure know,
> To Grecian gardens their extraction owe,
> One species is with robes of silver born,

The gen'rous scarlet by another worn,
And both the spring with early pride adorn.
Corsu and Coritus with both abound,
And much of each in shady Zacynth found,
Thousands in summer shine with either dye,
And in autumnal months they multiply.

Caroline Anne Southey wrote "The Sweet-scented Cyclamen" (1854), in which she empathized with the timidity and fragility of the flower:

I love thee well, my dainty flower!
My wee, white, cowering thing.
That shrinketh like a cottage maid,
Of bold, uncivil eyes afraid,
Within thy leafy ring!

Perhaps the loveliest of tributes to the cyclamen was penned by Mrs. S. O. Beeton in *All About Gardening* (1865). She noted that the flowers are "Most beautiful, graceful, and ladylike; so easily cultivated withal, that anyone may enjoy these floral bijous, either in the sitting room window, conservatory or greenhouse."

In my North Carolina garden, the parade of white, pink, and carmine-colored hardy cyclamen lasts from autumn until spring. I am continually amazed at the seedlings that pop up along paths, the result of ants lugging the seed to their stores. I am captivated by the bright petals hovering like helicopters over dead winter leaves in the afternoon sun, giving me an appreciation for "the fairer flowers" that Erasmus Darwin called "The gentle cyclamen, with dewey eye."

DAFFODIL

We weep to see you haste away

One of the best-known spring flowers is the daffodil, which is taxonomically the *Narcissus*, a member of the Amaryllidaceae. All its approximately fifty bulbous species are native to the Northern Hemisphere and currently fall within twelve distinct sections of the genus and include scores of cultivated forms. Species that are well known in gardens include *N. bulbocodium*, the hoop petticoat daffodil; *N. poeticus*, poet's narcissus or pheasant's eye; and *N. papyraceus*, paper-white narcissus. The many fine modern hybrid narcissi, including some miniatures, have gained wide garden acceptance.

Linnaeus chose the name *Narcissus* from Greek mythology. Echo, a nymph, was smitten with the beautiful youth Narcissus, but he rejected her, saying he cared for no woman's love. Upset at the rebuff, she enlisted the help of Cupid, who caused Narcissus to fall in love with his own image in a reflecting pool. With his love ever unrequited, he pined and wasted away beside the pool. While the nymphs prepared his funeral pyre, his body was transformed into the flower we now call narcissus. In one telling of the story, he drowned in the pool and reappeared as a flower growing along the River Styx. Contrary to the message sent by the myth, the language of flowers assigned to daffodils the connotations of chivalry and regard. According to Pliny and Virgil, the name narcissus derives from *narke*, meaning stupor or "narcotic effect," supposedly from the intoxicating fragrance of the flowers. Thus the flower name probably preceded the legend of Echo and Narcissus.

Besides the common name of daffodil, the plant is sometimes called jonquil in the American Southeast, which comes from *juncus*, the Latin word for a rush plant, because of the rush-like leaves; however, the jonquil is properly *Narcissus jonquilla*. The name daffodil appears to share its origin with the name asphodel, but the latter plant is a member of the Liliaceae and is not closely related to the daffodil (see the chapter "Asphodel"). William Turner suggested in *The Names of Herbes* (1548) the relationship between the names daffodil and asphodel: "Asphodillus groweth . . . in gardines in Antwerp, it maye be named in Englishe whyte affodil or Duche daffodil." The interchanging of the names is further confirmed in Henry Lyte's translation of Rembert Dodoens's *Niewe Herball or Historie of Plantes* of 1578: "This herbe [Asphodelus] is called . . . in English also Affodyl, and Daffodyll."

Robert Herrick wrote of the daffodil in "Divination by a Daffadill," finding in the flower the sign of his own life's inevitable course:

> When a Daffodill I see
> Hanging down her head t'wards me,
> Guess I may what I must be:
> First, I shall decline my head;
> Secondly, I shall be dead;
> Lastly, safely buried.

In "To Daffadills" (1648), Herrick wrote,

> Faire daffadills, we weep to see
> You haste away so soone;
> As yet the early-rising sun
> Has not attain'd his noone.

In *The Winter's Tale* by Shakespeare, Autolycus, a balladeer and rogue, sings of the English countryside on a road to Old Shepherd's cottage:

> When daffodils begin to peer,
> With heigh! the doxy over the dale,
> Why, then comes in the sweet o' the year;
> For the red blood reigns in the winter's pale.

A famous praise of the flower is by Wordsworth in "Daffodils":

> I wander'd lonely as a cloud
> That floats on high o'er vales and hills,
> When all at once I saw a crowd—
> A host of golden daffodils.
> Beside the lake, beneath the trees
> Flutt'ring and dancing in the breeze.

Aubrey De Vere sang the highest praise in his "Ode to the Daffodil," an early poem in which he called the flower the "lone-star of the unbeloved March":

> Herald and harbinger! With thee
> Begins the year's great jubilee!
> Of her solemnities sublime
> A sacristan whose gusty taper
> Flashes through earliest morning vapour,
> Thou ring'st dark nocturns and dim prime.

The myth of Narcissus and Echo has been told by many. Milton wrote of them in *Comus* (1627), a masque about a pagan disguised as a shepherd:

> Sweet Echo, sweetest nymph that liv'st unseen
> Within thy airy shell
> By slow Meander's margent green,
> And in the violet embroider'd vale

> Where the lovelorn nightingale
> Nightly to thee her sad song mourneth well;
> Canst thou not tell me of a gentle pair
> That likest thy Narcissus are?

Joseph Addison, following Ovid, described the naiads and dryads mourning the death of the youth in "Death of Narcissus": "And now the sister nymphs prepare his urn;/When, looking for his corpse, they only found/A rising stalk, with yellow blossoms crown'd."

John Keats described the daffodil simply in *Endymion* (1818):

> A thing of beauty is a joy forever.
> Its loveliness increases. It will never
> Pass into nothingness.
>
> .   .   .   .
>
> Such the sun, the moon.
> Trees old and young; sprouting a shady boon
> For simple sheep; and such are daffodils
> With the green world they live in.

And Oscar Wilde referred to narcissus in *The Picture of Dorian Gray* (1890), his novel of an English dandy who manages to escape aging while his likeness in his portrait grows older: "This young Adonis, who looks as if he was made of ivory and rose-leaves. Why, my dear Basil, he is a Narcissus."

DAHLIA

In colour as bright as your cheek

Dahlias are natives of Central America, Mexico, and Colombia and are members of the Compositae (Asteraceae). This important perennial ornamental garden plant has approximately thirty species, all with daisy-like heads and rays. As many as 20,000 cultivars exist—some spectacularly doubled—that no longer resemble the wild species, which is a simple single flower with a scarlet ray and yellow disk. *Dahlia pinnata* (*D. variabilis*) is one of the parents of numerous named cultivars. The tree dahlia, *D. imperialis*, is a large shrub to 9 meters (30 feet) high with lavender ray flowers.

Dahlias are named after the Swedish botanist Andreas Dahl, a student of Linnaeus. In eastern Europe and Russia, the dahlia is known as georgina, named for a St. Petersburg botany professor, Johann Georgi, who was so delighted with the dahlias he saw on a holiday in Paris that he took them home and gave them to friends. In the language of flowers a single dahlia indicated good taste, but a collection noted instability, apparently referring to the difficulty of growing it in cooler European gardens.

Dahlias were discovered by Europeans during the Spanish conquest of Mexico early in the sixteenth century. Dahlias were cultivated by the Aztecs primarily for their tuberous roots, which were used for medical purposes. Francisco Hernandez, a botanist to King Philip II of Spain, described dahlias in *The History of Mexico*, a book posthumously published in 1651. However, the plants themselves did not reach European gardens until the late eighteenth century, coming first to Madrid, then to Paris where Empress Josephine is said to have planted the first tubers herself at Malmaison, and finally to London. A species of *Dahlia* was listed in the earliest-known herbal produced in North America, the *Badianus Herbal*. It was compiled in Latin in 1552 by sons of Aztec noblemen who were pupils at the Roman Catholic College at Santa Cruz, Mexico. The colorful herbal was rediscovered at the Vatican Library in 1931 and reproduced in facsimile and in translation in 1940.

In Aztec mythology, the native dahlia, known as *cocoxochitl*, is a symbol of the Serpent Woman, who daily conversed with an eagle on Serpent Mountain, near the sky gods. On one of her visits, when the eagle was relaying messages from the gods, she found a rabbit sitting beside an agave, holding in its mouth a dahlia with eight red flower rays. The Serpent Woman took the dahlia, and the gods told her to impale the flower on the sharp leaves of the agave and hold it to her breast all night long. The next morning the Serpent Woman delivered a full-grown son, Uitzilopochtli, the War God, who was fully armed with the sword-like leaf of the agave. The eight blood-red rays had given him strength for war and a thirst for blood. After that time, the Aztecs sacrificed prisoners to the War God every eight years, removing their hearts and placing them on stones surrounded by dahlias and agave.

English novelist Robert Smith Surtees in *Handley Cross* (1843) revealed his knowledge of how cold-tender dahlias are: "Hurrah! . . . it is a frost! The dahlias are all dead!" A more positive sentiment appears in a note Lord Holland sent to his wife, Lady Holland, in July 1824. She had seen dahlias growing in Madrid and sent seed back to England to see if Lord Holland's librarian (and gardener) could get them to grow. He succeeded to the delight of Lord Holland, who wrote the following doggerel to his wife:

The dahlia you brought to our isle
Your praises for ever shall speak;
Mid gardens as sweet as your smile,
And in colour as bright as your cheek.

English humorist Thomas Hood wrote in his satire *Miss Kilmansegg and Her Precious Leg* (1843), "And ask the gardener, Luke or John,/Of the beauty of double-blowing—/A double dahlia delights the eye." Longfellow wrote of "dahlias in the garden walk" in "Wayside Inn: Student's Tale" (1863). The dahlia has also inspired poet Martin, who composed "The Dahlia" to show the flower as an example to follow, ever bright and hopeful though put to hardship:

Though severed from its native clime,
Where skies are ever bright and clear,
And nature's face is all sublime,
And beauty clothes the fragrant air,
The dahlia will each glory wear,
With tints as bright, and leaves as green;
And winter, in his savage mien,
May breathe forth storm—yet she will bear
With all:—and in the summer ray,
With blossoms deck the brow of day.

And thus the soul—if fortune cast
Its lot to live in scenes less bright,—
Should bloom amidst the adverse blast;—
Nor suffer sorrow's clouds to blight
Its outward beauty—inward light.
Thus should she live and flourish still,
Though misery's frost might strive to kill
The germ of hope within her quite:—
Thus should she hold each beauty fast,
And bud and blossom to the last.

Edith Matilda Thomas in "Frost To-Night" remembered the child-hood event of gathering the dahlias on the last evening of the season: the anticipation, the sadness, and the pride, sensations that she felt again in her waning years.

Apple-green west and an orange bar,
And the crystal eye of a lone, one star . . .
And, "Child, take the shears and cut what you will.
Frost to-night—so close and dead-still."

Then, I sally forth, half sad, half proud,
And I come to the velvet, imperial crowd,
The wine-red, the gold, the crimson, the pied,—
The dahlias that reign by the garden-side.

The dahlias I might not touch till to-night!
A gleam of the shears in the fading light,
And I gathered them all,—the splendid throng,
And in one great sheaf I bore them along.

In my garden of Life with its all-late flowers
I heed a Voice in the shrinking hours:
"Frost to-night—so clear and dead-still . . ."
Half sad, half proud, my arms I fill.

DAISY

Show'd like an April daisy on the grass

What a daisy is to the gardener may differ from what it is to the botanist. Many plants are called daisies, and unraveling their identities can prove taxing. To the English and continental gardener the most common daisy is *Bellis perennis*, whose genus name means pretty in Latin. The Marguerite daisy of Europe refers generally to any large-flowered daisy. It was originally *Leucanthemum vulgare*, but this species has many additional common names, including moon daisy, moon pennies (in Yorkshire), dog daisy, and ox-eye daisy. The name Marguerite comes to the English language from French and Spanish and means pearl. Well known to North American gardeners are the shasta daisies, formerly *Chrysanthemum maximum* but now *Leucanthemum ×superbum*. Other flowers called daisies include the African (*Arctotis* and *Dimorphotheca*), the Transvaal or Barberton of South Africa (*Gerbera*), Michaelmas (*Aster*), and *Chrysanthemum*. Creating a stir among gardeners was the renaming of chrysanthemums of gardens and the florist trade as *Dendranthema ×grandiflora*, followed by the change back to *Chrysanthemum*. Regardless of the taxonomy, all these are daisies and are members of the vast family Compositae (Asteraceae), distributed worldwide.

After the rose and the lily, the daisy is probably the flower most often mentioned by poets. Chaucer was passionate for it, Shakespeare included daisies in the drowning Ophelia's "fantastic garland" of flowers, and Wordsworth wrote at least three poems to it. Robert Burns, Ben Jonson, Percy Bysshe Shelley, John Clare, and countless others referred to this simple but adored flower. Its popularity during the Victorian era was reflected in the language of flowers with as many sentiments as there are types of daisies. For example, the red daisy meant unconscious thoughts, whereas the white daisy suggested innocence, and the wild daisy of the fields communicated, "I shall think about it." And an old saying claims that summer has come when you set foot on six (or seven) daisies at once.

During the days of chivalry and knighthood, daisies were worn by suitors and included as designs on mementos for waiting ladies. Daisies became emblems for various celebrations, including those associated with church calendar days. The midsummer daisy (*Bellis perennis*) was the flower used in English church decorations on St. Barnabas Day, 11 June. The "lucken gowan" and other wild field daisies were used on 6 May, the day of St. John the Evangelist Before the Latin Gate. The Latin Gate was one of the entrances to Rome at which St. John was ordered to be boiled in a cauldron of oil. The Marguerite daisy (*Leucanthemum vulgare*) was the floral emblem of St. Margaret, the supposed mother of two kings of Scotland, whose feast day is 20 July. It used to be called herb Margaret, to which Chaucer referred: "The daisie, a floure white and rede,/In frenche called la belle Marguerite."

From early on the daisy had dual uses as an oracle and as an herb. It became popular for telling prophesies because its composite form, or collection of numerous florets comprising the flower head, made the petals easy to pluck. Goethe supposedly gave Marguerite a daisy-like flower to determine Faust's love for her as she recited the now popular chant, "He loves me. He loves me not." Further, because the flower head looks much like an eye, herbalists using the ancient Doctrine of Signatures thought daisies could cure bloodshot eyes and various other eye problems.

According to classical mythology, the daisy obtained its name from a dryad who was the granddaughter of Danaeus, the powerful king of Argos who had fifty daughters known as the danaeids. Engaged to Epigeus, the deity of rural areas, she one day attracted the attention of Vertumnus, the guardian deity of orchards. To escape his amorous pursuit, the young tree nymph appealed to the gods. They transformed her into a humble flower and gave her the name Bellis. The legend was commemorated by René Rapin:

> When the bright ram, bedecked with stars and gold
> Displays his fleece, the daisy will unfold,
> To nymphs a chaplet, and to beds a grace,
> Who once her self had born a virgin face.

Chaucer mentioned the daisy often in his poetry. To him the flower was the "eye of the day" or "day's-eye," because he thought the flower closed at night after having displayed its face during the day. In England Chaucer would likely have known *Bellis perennis*; however, he had ample opportunity to observe the daisy and its other kin during his travels to Italy and France on official business for Edward III. The trips were so successful diplomatically that in 1367 he was granted a life's pension, including a daily pitcher of wine from the king's butler.

In *The Legend of Good Women*, an unfinished work composed between 1372 and 1386, after his pension began, Chaucer used the daisy several times to celebrate women for their fidelity:

> And she was clad in real habit grene,
> A fret of gold she hadde next hir heer,
> And upon that a whyt coroun she beer
> With florouns smale, and I shal not lye
> For al the world, ryght as a dayesye.

He must have been familiar with Euripides's Greek tragedy *Alcestis* (438 BCE), in which Queen Alcestis saves the life of her husband, King Admetus, by agreeing to descend into Hades in his place. She in turn is spared by the efforts of Heracles, or Hercules. But Chaucer, perhaps borrowing from other legends of fantastic transformations, changed the Euripides story by transforming the heroine into a daisy. In *The Legend of Good Women*, Chaucer interchanged the name Alceste and the title Queen of Love for a character representing Queen Anne of Bohemia, who married Richard II, the grandson of Edward III. In the last section of the tale, Alceste is transformed into a daisy with as many virtues as there are petals on the flower head.

Further evidence of the popularity of the daisy appears in literature of the 1500s and 1600s. In *The Shepheardes Calender* (1579) by Edmund Spenser, the June entry describes a shady glade and grassy ground with "daintye daysies" growing amid brambles and birds of every kind. Robert Herrick wrote tenderly of his Julia in "To Daisies, not to shut so soon," comparing the closing of the flower to the loss of his mistress:

> Shut not so soon; the dull-eyed night
> Has not as yet begun
> To make a seizure on the light,
> Or to seal up the sun.

Alexander Pope mentioned daisies in his celebration of the vernal season, "Spring, the First Pastoral." The shepherds Daphnis, Strephon, and Damon are enjoying a spring day on "Windsor's blissful plains" on the banks of the Thames. Daphnis admires the bird songs and asks Strephon and Damon to sing. Damon replies,

> Then sing by turns, by turns the Muses sing,
> Now hawthorns blossom, now the daisies spring,
> Now leaves the tree, and flow'rs adorn the ground;
> Begin, the vales shall ev'ry note rebound.

Shakespeare did not overlook the daisy in his writings. In *Hamlet* (1601) Ophelia says to her brother, Laertes, "There's a daisy; I would give you violets, but they withered all when my father died: they say he made a good end." In *The Rape of Lucrece* Shakespeare compared the heroine's complexion to the white of a daisy:

> Without the bed her other faire hand was
> On the green coverlet; whose perfect white
> Show'd like an April daisy on the grass,
> With pearly sweat, resembling dew on the night.

And in *Love's Labour's Lost* (1598), "The Song of Spring" that ends the play includes daisies in the signs of the season:

> When daisies pied and violets blue
> And lady-smocks all silver-white
> And cuckoo-buds of yellow hue
> Do paint the meadows with delight,
> The cuckoo then, on every tree,
> Mocks married men.

Shakespeare, ca. 1610 in *Cymbeline*, again used daisies as a sign, but this time to represent something more sinister:

> Let us
> Find out the prettiest daisied plot we can,
> And make him with our pikes and partisans,
> A grave: come, arm him.

William Wordsworth, an early English romantic, is well known for his poems that idealize nature. The following selection from "To the Daisy" is, like his more famous "The Daffodils," an example of English romantic poetry of the early nineteenth century in which nature is appreciated for its own sake:

> With little here to do or see

> Of things that in the great world be,
> Sweet daisy! oft I talk to thee.
> For thou art worthy,
> Though unassuming commonplace
> Of nature, with that homely face,
> And yet with something of a grace,
> Which love makes for thee!

In "A Farewell" Wordsworth says adieu to his little "nook of mountain ground." Among the characters he will miss is the daisy, which he calls the gowan; however, he appears to have described a *Ranunculus*:

> We leave you here in solitude to dwell
> With these our latest gifts of tender thought;
> Thou, like the morning, in thy saffron coat,
> Bright gowan, and marsh-marigold, farewell!
> Whom from the borders of the Lake we brought,
> And placed together near our rocky Well.

The botanical identity of the gowan varies from region to region, but the linguistic origin of the name is Scottish, in which it means fair. The gowan is sometimes thought of as any field flower, occasionally as a globe flower (*Trollius*), and variously as both a globe daisy (*Globularia*) and field daisy (*Bellis*).

Charles Dickens refers to the gowan in *David Copperfield* (1849):

> "Then I will drink," said Mr. Micawber, "if my friend
> Copperfield will permit me to take that social liberty, to
> the days when my friend Copperfield and myself were
> younger, and fought our way in the world side by side. I
> may say, of myself and Copperfield, in words we have
> sung together before now, that
>   'We twa hae run about the braes,
>   And pu'd the gowans fine'.

—in a figurative point of view—on several occasions. I am not exactly aware," said Mr. Micawber, with the old roll in his voice and the old indescribable air of saying something genteel, "what gowans may be, but I have no doubt that Copperfield and myself would frequently have taken a pull at them, had it been feasible."

Dickens was borrowing Mr. Micawber's song from Robert Burns, who thought of his childhood days and friends in "Auld Lang Syne" (1788):

> We twa hae run about the braes,
> And pou'd the gowans fine;
> But we've wandered monie a weary fit
> Sin' auld lang syne.

Burns once inadvertently plowed under a daisy and apologized in "To a Mountain Daisy" (1786):

> Wee, modest, crimson-tippèd flow'r,
> Thou's met me in an evil hour,
> For I maun crush amang the Stoure
> Thy slender stem;
> To spare thee now is past my pow'r,
> Thou bonnie gem.

John Clare wrote of the daisy several times. In "The Eternity of Nature" (ca. 1828) he steps on a daisy only to have it provoke the thought of a daisy's perspective: "Trampled under foot,/The daisy lives, and strikes its little root/Into the lap of time."

From childhood onward, we easily recognize the daisy as a flower of the field and roadsides with its simple, often overlooked beauty. Though some may quarrel regarding the flower's taxonomy and nomenclature, those reared on English literature may revel in Words-

worth's line from one of his four poems titled "To the Daisy"—
"Bright flower whose home is everywhere"—or in Coventry Patmore's
elegant yet simple couplet, "The daisies coming out at dawn/In con-
stellations on the lawn." But we may also empathize with John Keats,
who said on his deathbed that he could already feel the daisies growing on his grave, a feeling that survives today in the expression "pushing up daisies" from one's grave.

DAYLILY

Whene'er the kettle bubbles

Daylilies are members of the Liliaceae and comprise fifteen species that make up the genus *Hemerocallis*. The genus name is derived from the Greek words *hemera*, for day, and *kallos*, for beauty. Bishop Joseph Hall wrote that "the hemerocallis is the least esteemed because one day ends its beauty." René Rapin contemplated the short-lived beauty of the flower:

> And those whose blossoms curl
>   obliquely back,
> Rib'd on the sides with a bright scar-
>   let streak,
> Shalt of day-lily the fair name receive,
> If one whose summer's day the beautys
>   live,
> These into garlands may the virgins
>   twine,
> When fresh and plenteous on the beds
>   they shine.

In the language of flowers, daylilies suggested coquetry. An Old English name was *lilly-asphodill*, because many thought the plant was a combination of the lily and the asphodel; the name survives in the currently recognized species *Hemerocallis lilio-asphodelus* (formerly *H. flava*).

Long before they traveled to Europe, daylilies were cultivated in China where they were used as food coloring, and the flower buds were added to soups. In China they were called the plants of forgetfulness in the belief that the flowers would cure sorrow by inducing memory loss. In Korea the daylily is called forget-your-troubles. A quatrain, possibly referring to *Hemerocallis middendorffii*, suggests various uses for this yellow species:

> This lily is Korea's cure
> For everybody's troubles;
> Its leaves as food bring heirs for sure,
> Whene'er the kettle bubbles.

EUPHORBIA

Three cups in one

Euphorbia, or spurge, consists of about 2000 species distributed world-wide. They may be annuals, perennials, shrubs, or trees, and many are succulent. Most of the species have diminutive flowers and look like cacti. Many members of the Euphorbiaceae contain a milky white latex that may sting or burn sensitive skin. A common garden species is *E. lathyris*, the caper spurge or mole plant, which has long been believed to repel moles. The name caper spurge refers to the use of the fruit—the seed is poisonous—as a poor man's substitute for capers (*Capparis spinosa*). *Euphorbia milii* (formerly *E. splendens*), the crown of thorns plant, is not frost tolerant and is a common houseplant. Popular at yuletide is the showy *E. pulcherrima*, the poinsettia. And a well-known North American species is *E. marginata*, or snow on the mountain.

The name spurge is derived from Old French *espurgier*, meaning to purge, an allusion to the use of the milky latex as a purgative for cleansing and purifying the body. The herbalist William Turner observed the medicinal nature of euphorbia in 1562: "Spurge purgeth thynne fleme vehemently." The milky substance from the euphorbias was used for medicinal purposes by Euphorbus, a physician to King Juba II of Mauretania and Numidia during the first century CE. At the same time, spurge was known to Pliny the Elder, who noted that in Greek and Latin the name *euphorbium* was applied to both the plant and its sticky juice. When Linnaeus classified plants in the 1700s using a binomial system, he retained the classical name for the genus.

John Gerard, in *The Herbal or General History of Plants* (1633 revision by Thomas Johnson), described the properties of the "gummie euphorbium," borrowing, he says, from Pliny and other writers:

> Euphorbium (that is to say, the congealed juice which we use) is of a very hot, and as Galen testifieth, caustic or burning facultie, and of thinne parts: it is also hot and dry in the fourth degree . . . [and] if it be inwardly taken it purgeth by siege water and flegme; but withall it setteth on fire, scortcheth, and fretteth, not only the throat and mouth, but also the stomacke, liver, and the rest of the intrals, and inflames the whole bodie.

Chaucer knew the medical properties of spurge. In *The Nun's Priest's Tale*, spurge is a laxative that the cock Chauntecleer's hen wife recommends to rid him of indigestion and, hence, his nightmares. A different use of spurge was noted by Englishman Benjamin Stillingfleet in 1762 in *Miscellaneous Tracts Relating to Natural History, Husbandry, and Physick*: "The spurge, that is noxious to man, is a most wholesome nourishment to the caterpillar." And Alfred, Lord Tennyson, in "Last Town" (1872) noted "That he can make figs of out thistles [and] milk from burning spurge."

Spurge was also appreciated simply as an attractive plant. Henry David Thoreau wrote of the coastal euphorbia (*Euphorbia polygonifolia*) in *Cape Cod* (1865): "The plants which I noticed here and there on the pure sandy shelf, between the ordinary high-water mark and the foot of the bank . . . were Sea Rocket, . . . Saltwort, . . . Seaside Spurge." The small flowers of *E. amygdaloides* were remembered by Dante Gabriel Rossetti in "The Woodspurge":

My eyes, wide open, had the run
Of some ten weeds to fix upon;
Among those few, out of the sun,
The woodspurge flowered, three cups in one.

From perfect grief there need not be
Wisdom or even memory:
One thing then learnt remains to me—
The woodspurge has a cup of three.

EVENING PRIMROSE

Blooms on while night is by

The genus *Oenothera* contains about 125 species referred to variously as evening primrose, sundrops, or suncups. All are native to North and South America and are members of the Onagraceae; they are unrelated to the true primrose, *Primula*, of the family Primulaceae. In the 1600s, evening primrose made its way to European gardens, initially to Padua in northern Italy, via seed sent from the American colonies, most likely Virginia. In England, the plant naturalized and became a common wild flower; now the evening primrose is generally regarded as a member of the British flora. The common evening primrose is *Oenothera biennis* with soft sulfur-yellow petals that open at dusk. Sundrops, *O. fruticosa*, with deeper yellow petals, opens during the day. Another species commonly found in gardens is the Ozark evening primrose, now *O. macrocarpa* but formerly *O. missouriensis*, with lemon chalices. Two of the showiest are *O. caespitosa*, endemic to the Rocky Mountains, and *O. rosea* of the southern plains of the United States.

Linnaeus took the genus name from classic Greek, though *Oenothera* had been attributed to a different plant, probably *Oleander* or *Epilobium*. Latinized into *oinos thera*, the name means "the hunt for wine," which Linnaeus probably applied after the influence of Theophrastus, who thought the original oenothera roots gave off an odor reminiscent of wine. Though the American oenothera for which Linnaeus appropriated the name could not have been known by Theophrastus, its roots are edible. When the plant reached England it was given the common name of the "tree primrose of Virginia" by John Parkinson, because the flowers, he thought, smelled like the native English primrose and the plant grew tall—up to 1.5 meters (5 feet) or so. John Gerard's *Herbal or General History of Plants* (1633 revision by Thomas Johnson) maintained Parkinson's name and noted that "this Virginian hath beene described and figured . . . by Mr. Parkinson by the name of *Lysimachia lutea*," thinking, too, that the plant was a willow-herb in the primula family. Gerard went on to say that "these floures have somewhat the smell of a primrose; after the floures are fallen, the cods grow to be some two inches long. It floures in June, and ripens the seed in August."

Hilderic Friend, an English parish priest and the author of *Flowers and Flower Lore* (1884), wrote about the evening primrose:

> Our flower is the Evening Primrose, the latter portion of the name being derived from the pale, primrose tint of the blossoms; the former, from its beginning to wake up just as other flowers are going to sleep. Here and there a blossom may sometime be seen expanded in the daytime, but the majority of flowers do not open till six or seven o'clock in the evening, and then they are slightly fragrant, a beautiful characteristic of many night-blooming flowers. It would seem that, as they open during a period when beauty of appearance would be disregarded, on account of darkness, they make up for the disadvantage by the diffusion of the choicest odours. . . .
>
> The Evening Primrose is well known from its regularly opening about the time of sunset, and shutting again with a loud popping noise about sunrise. After six o'clock these flowers regularly report the approach of night.

John Langhorne, a rector at Blagdon, Somerset, in the middle of the 1700s, wrote several volumes of poetry, among them *Fables of Flora*, which includes "The Evening Primrose":

Than vainer flowers, though sweeter far,
The Evening Primrose shuns the day;
Blooms only to the western star,
And love its solitary ray.

In Eden's vale an aged hind
At the dim twilight's closing hour,
On his time-smoothed staff reclined,
With wonder viewed the opening flower.

John Keats was also moved by the opening of the evening primroses:

A tuft of evening primroses,
O'er which the mind may hover till it dozes;
O'er which it well might take a pleasant sleep,
But that 't is ever startled by the leap
Of buds into ripe flowers.

Felicia Dorothea Hemans, born in Liverpool, published fourteen volumes of verse in the early 1800s. She was popular in the United States and is now seen as a vanguard, pre-Victorian female poet who paved the way for a succession of women poets. She wrote the following verses in an allusion to the "dial of flowers," a schedule that Linnaeus created describing the daily folding and unfolding of flowers:

'Twas a lovely thought to mark the hours
As they floated in light away,
By the opening and the folding flowers,
As they laugh to the summer's day.

Thus had each moment its own rich hue,
And its graceful cup and bell,
In whose coloured vase might sleep the dew,
Like a pearl in an ocean shell.

In the Victorian language of flowers the evening primrose was a symbol of silent love, which might have arisen from the flowers' subtle habit of not opening during the day. It was this habit that inspired John Clare to characterize the evening primrose as delicate and shy:

> When once the sun sinks in the west,
> And dewdrops pearl the Evening's breast;
> Almost as pale as moonbeams are,
> Or its companionable star,
> The Evening Primrose opes anew
> Its delicate blossoms to the dew;
> And, hermit-like, shunning the light,
> Wastes its fair bloom upon the Night;
> Who, blindfold to its fond caresses,
> Knows not the beauty he possesses;
> Thus it blooms on while Night is by;
> When Day looks out with open eye,
> Bashed at the gaze it cannot shun,
> It faints, and withers, and is gone.

FERN

Where the brackens make their bed

Ferns are primitive vascular plants that do not produce flowers. They reproduce by tiny spores usually borne on the underside of their leaves or fronds. Ferns belong to a major division of the plant kingdom called Pteridophyta; true ferns comprise nearly 240 genera and at least 10,000 species worldwide. Whether indoors or out, ferns have been grown in gardens and conservatories for centuries.

The emerging frond of a fern resembles the neck of a fiddle and is thus called a fiddlehead or crozier. In medieval times, the fiddlehead was called St. John's hands or lucky hands, and those who carried it in their pockets were said to be protected from witches. Fronds were often placed over doorways to ward off lightning. The Cherokee Indians noticed the uncoiling of the fiddlehead and believed that the plant had "uncurling properties" that would cure tightened arthritic limbs.

Though ferns do not reproduce by seed, for centuries people in Great Britain believed that ferns produced invisible seed, because they could not find any. According to the Doctrine of Signatures, a seventeenth-century theory that assigned medicinal properties to plants based on their physical resemblance to parts of the human body, those who possessed the secret of the invisible seed were themselves also invisible and imbued with magical charms. In Shakespeare's *Henry IV, Part I* (1598), Gadshill, a crony of the dissolute Prince Hal, has been sent to an inn by the prince to identify travelers who could be robbed for sport. Gadshill says to the chamberlain, a co-conspirator, "We steal as in a castle, cock-sure: we have the receipt of fern-seed, we walk invisible." To which the chamberlain responds, "Nay, by my faith, I think you are more beholding to the night than to fern-seed for your walking invisible."

The magical and mystical power of fern seed also was a part of the celebration of midsummer night, or the summer solstice, 21 June. That night is also called St. John's Eve in honor of his nativity, which long ago was set as 24 June to coincide with the presumed, but astronomically incorrect, date of the summer solstice. Legends say that at midnight, the precise moment of St. John's birth, the fern blooms quickly and produces vast quantities of invisible seed. On that night, a stack of twelve pewter plates, representing the twelve apostles of Jesus, is placed underneath a stand of ferns, and as the seed falls from the fronds, it passes through the plates but rests invisibly on one. Which plate holds the seed is known only by those knowing the "receipt" or recipe. The devil is obliged to bring to the person who catches the invisible seed treasures and magical powers, such as to ward off evil or to open locks, according to various tellings of the legend. The legend of the invisible seed is noted in this anonymous quatrain:

> But on St. John's mysterious night,
> Sacred to many a wizard spell,
> The hour when first to human sight,
> Confest, the mystic fern-seed fell.

The lady fern is *Athyrium filix-femina*. The generic name is obscure but possibly from the Greek *athoros*, for breeding well, in reference to diverse forms of its sori, the clusters of spores on the underside of a frond. The specific epithet is the Latin translation of "lady fern." The fern was described by Major Robert Calder Campbell. He was an 1839 retiree of the Indian army who wrote and was active in literary circles, encouraging Dante Gabriel Rossetti's writing efforts. Of the lady fern he wrote,

> If you would see the lady fern
> In all her graceful power,
> Go look for her where woodlarks learn
> Love-songs in a summer bower.
>
> . . . .
>
> While harts' tongues quiet with a green more bright
> Where the brackens make their bed.
> Ferns all—and lovely all—yet each
> Yielding in charms her
> Natural graces Art might teach
> High lessons to confer.

The adder's tongue fern—*Ophioglossum vulgatum*, whose genus name means snake tongue, a reference to the fertile frond within the sterile leaf blade—was used in witchcraft during incantations and the casting of spells. Legend says that if this fern is wrapped in new wax and put in the left ear of a horse, it will make the horse fall down; when the wax is removed the animal will rise. Adder's tongue fern was used in the pharmacopeia to treat snakebites and disorders of the tongue, in accordance with the shape of the fern and the Doctrine of Signatures.

The dried bracken fern—*Pteridium*, whose name means lacy and feathery, an allusion to the shape of the leaflets—was used as a source of alkali. The name bracken is related to the words brake and brack, which mean uncultivated ground or underwood, the areas where bracken grows. The brake fern is *Pteris*, which name means wing-like. An old English saying is "When the fern goes from brown to red,/ then milk is good with brown bread," suggesting that hillsides rich with autumn bracken provided full pastures for grazing cattle. One anonymous poet saw bracken more negatively, however, using it to show the waste of years:

> Thus is my Summer worne away and wasted,
> Thus is my harvest hastened all to rathe;
> The eare that budded faire is burnt and blasted
> And all my hopéd graine is turned to scathe.
> Of all the seede that in my youth was sowne
> Was not but Brakes and brambles to be mowne.

According to legend, the maidenhair fern *Adiantum* is named in reference to Aphrodite, or Venus in Roman mythology. She is supposed to have come from the sea, rising from the water with dry, flowing hair that was made from the fern *Adiantum*. And indeed, the generic name means dry. One species, *A. capillus-veneris*, has a specific epithet meaning Venus's hair.

In the language of flowers, ferns suggested sincerity. The fern is recommended for church decorations on St. Matthias Day, 24 February. Saint Matthias was one of the disciples of Jesus, but any legendary link between him and the fern is unclear.

FORGET-ME-NOT

Hope's gentle gem

The family Boraginaceae includes about fifty species of *Myosotis*. They are distributed primarily in Europe; however, disjunct species are native to temperate regions in North America, New Zealand, eastern Africa, and Asia. The familiar forget-me-not of gardens is *M. sylvatica*, whose numerous cultivars, such as 'Blue Bird' and 'Blue Basket', are perennial but are often treated as biennials. Most species have blue flowers with yellow centers. The water forget-me-not, *M. palustris*, flourishes along river banks, on water's edge, and in moist places. Two forget-me-nots are alpine, *M. alpina* and *M. alpestris*.

Myosotis is from Greek and means mouse (*mus*) ear (*otos*), a reference to the small pointed leaves. Often the plant is simply called mouse ear. The more common name forget-me-not is the same in English, French (*ne-m'oubliez-pas*), German (*Vergiss-meinnicht*), and other European languages. An older common name is scorpion-grass, because of a resemblance between the curled raceme of the unopened flowers and a scorpion's tail poised to strike. For this reason *M. scorpioides* is considered by the Doctrine of Signatures to be an antidote for stings and bites from bees, wasps, and scorpions. An old common name *herba-clavorum*, from the Latin for plant and nail, came from blacksmiths' practice of twining mouse ear through horses' manes to reduce their pain while shoeing them.

Circa 1390 Henry of Lancaster, who became Henry IV of England, selected a form of the *sou-*

veigne vous de moy, the "remember me" flower, as an emblem for his robes and collars, signifying true love. By the fifteenth century, the forget-me-not ensured that those wearing it should never be forgotten by their lovers. A rather charming old name applied to forget-me-not in Yorkshire is think-me-on.

An old legend gives the story of the plant's common name. In the Garden of Eden, as God (or Adam according to some versions) named each plant, He cautioned it not to forget its name. As He turned to leave the garden, a small voice asked by what name it should be called because it had been overlooked. God responded, "Because I forgot you before, and so that I shall not forget again, you shall be called forget-me-not." Thus, in the language of flowers, it is generally associated with remembrance as well as with true love. The story is told with slight alteration in this anonymous rhyme:

> When to the flowers so beautiful
> The Father gave a name,
> Back came a little blue-eyed one
> (All timidly it came);
> And standing at its Father's feet,
> And gazing in His face,
> It said in low and trembling tones,
> "Dear God, the name Thou gavest me,
> Alas! I have forgot."
> Then kindly looked the Father down,
> And said, "Forget Me Not."

Many European stories of the naming of the forget-me-not relate to a loved one's drowning in the Rhine, the Danube, or one of various Italian rivers. In a German legend, a knight and his lady were strolling along the banks of the Danube when they saw a pretty spray of blue flowers floating on the water but about to sink. The lady admired their delicate beauty and regretted their destiny; at this hint the knight plunged in to retrieve the flowers. The river current was too strong for him, but he flung the flowers to his bride, crying as he was swept away, "*Vergiss mich nicht.*" Here is one poet's ending to the story:

> And the lady fair of the knight so true,
> Aye remembered his hapless lot;
> And she cherished the flowers of brilliant hue,
> And braided her hair with the blossoms blue,
> And she called it Forget-me-not.

Miss Pickersgill expanded upon the story of the lady and her knight in "The Bride of the Danube":

> Alas! her tears, her sorrows now were vain,
> For him she loved she ne'er shall see again!
> Is this then a bridal, where, sad in her bower,
> The maid weeps alone at the nuptial hour;
> Where hushed is the harp, and silent the lute—
> Ah! why should their trilling strains be mute?
> And where is young Rodolph? where stays the bride groom?
> Go, ask the dark waters, for there is his tomb.
>
> Often at eve when maidens rove
> Beside the Danube's wave,
> They tell a tale of hapless love,
> And show young Rodolph's grave;
> And cull the flowers from that sweet spot,
> Still calling them "Forget-me-not."

In addition to appearing in numerous legends, forget-me-nots have been much noted in poetry and prose. But oddly, no forget-me-nots emerge among the floral references in Shakespeare's vast writings— rosemary is recommended by Perdita and Ophelia for remembrance, not forget-me-nots.

In "The Keepsake" (1817) from Samuel Taylor Coleridge's *Sibyll Leaves*, the narrator complains that the flowers have fallen from the rose:

Nor can I find, amid my lonely walk
By rivulet or spring, or wet roadside
That blue and bright-eyed flow'ret of the brook,
Hope's gentle gem, the sweet forget-me-not.

In "The Brook," Tennyson "move[s] the sweet forget-me-nots/That grow for happy lovers." And John Clare included the forget-me-not in a bouquet of flowers in "On May Morning":

The little blue Forget-me-not
Comes too on friendship's gentle plea,
Spring's messenger in every spot,
Smiling on all,—"Remember me!"

But in "A Bed of Forget-me-nots," Christina Georgina Rossetti takes a more cynical view of the flower and its significance: "Is love so prone to change and rot/We are fain to rear forget-me-not/By measure in a garden plot?"

In "Evangeline" (1847), Henry Wadsworth Longfellow used the flowers to describe the night sky: "Silently, one by one, in the infinite meadows of heaven,/Blossomed the lovely stars, the forget-me-nots of the angels." A North American species, *Myosotis verna*, has white petals, but the improbability of yellow forget-me-nots caused Cobham Brewer, editor of *The Dictionary of Phrase and Fable* (1894), to wryly comment under the entry "forget-me-nots of the angels": "The stars are so called by Longfellow. The similitude between a little light-blue flower and the yellow stars is very remote. Stars are more like buttercups than forget-me-nots."

Among the numerous legends associated with this charming little yellow-eyed blue flower—including the whispered messages of fidelity and love—few writers can exceed this anonymous tribute:

Of all the flowers that deck the field,
Or grace the garden of the heart,
Though others richer perfume yield,
The sweetest is forget-me-not.

FORSYTHIA

In the late spring the canaries come

The *Forsythia* shrub, a member of the Oleaceae, consists of six species and numerous hybrids. The genus is named after William Forsyth, a superintendent in the late eighteenth century of the royal gardens of Kensington Palace near London. At one time, crowns of the golden blossoms were placed by kings on the heads of scholars after they successfully passed the civil examinations that entitled them to official rank.

In Korea, the species *Forsythia viridissima* (formerly *F. koreana*) is called goldenbells, a reference to its bright yellow petals. According to legend, the forsythia marks the coming of springtime. Korean legend tells of another kind of rebirth heralded by forsythia—the rejuvenation of love. Once, a poet took a long journey and left his wife behind. On his return, he found his faithful wife waiting at the gate. Seeing her standing there, he appreciated her beauty for the first time. In her honor he wrote a poem that he sang to her:

> In the late spring the canaries come,
> The forsythia fades and the apricots fall;
> And in the bamboo shade of my mountain home
> Forever abides, my Love, my All.

FOXGLOVE

The glove of the wee people

The twenty species of foxglove are distributed naturally in Europe, northern Africa, and western Asia. They are biennial and perennial members of the Scrophulariaceae or figwort family. Foxgloves commonly found in gardens include *Digitalis purpurea*, with its purple-spotted throat; *D. lutea*, the yellow or straw foxglove; and *D. grandiflora*, with apricot-pink flowers, and its numerous cultivars. Less common are the rusty foxglove, *D. ferruginea*, and *D. lanata*, the Grecian foxglove of the Mediterranean. One species from the Balearic Islands off the coast of Spain is *D. dubia*, named because of the original doubt that it was a true foxglove.

The genus name, derived from the Latin word for finger, *digitus*, alludes to the flower's resemblance to a glove or finger. The plant was so named by Leonhard Fuchs in 1542. In 1629 John Parkinson wrote, "We call them generally foxglove in English; but some (as thinking it to be too foolish a name) do call them finger-flowers, because they are like unto the fingers of a glove, the ends cut off." The English name foxglove appears to be a corruption of folk's-glove, the glove of the fairies or wee people. But some suggest that such a correlation is silly. The plant is also called fairy-fingers, and regardless of the etymological origin, these quaint names richly provide the plant with legends. In some areas the plant is known as fox fingers, and associated stories suggest that the flowers are the gloves worn by foxes to keep dew off their paws.

To northern Europeans, however, the shape of the flowers did not suggest a glove or finger, but rather a bell. An old Norwegian legend refers to digitalis as fox-bell (*revbjölla*) in the belief that when foxes wore them, the eerie sound of the bell would scare away hunters who collected fox tails for good luck. The English name may derive from the Anglo-Saxon word *foxes-gleow*, *gleow* being the word for a musical instrument of rings of bells. The plant has also been variously known as witches' thimbles and fairy caps. The plant is known in French legends as *les gants de Notre Dame*, the gloves of the Virgin. In German, an old name for foxglove is *fingerhut*, for thimble.

A rhyme by William Browne of Tavistock maintained the thought of foxgloves as gloves, which were being sought by Pan for his mistress:

> To keep her slender fingers from the sunne
> Pan through the pastures oftentimes hath runne,
> To pluck the speckled foxglove from their stemme
> And on those fingers neatly placed them.

The flower made English poet Abraham Cowley think of gloves, too: "The foxglove on fair Flora's hand is worn/Lest while she gathers flowers she meets a thorn." But Christina Rossetti, taking a more realistic view, wrote in *Sing-Song: A Nursery Rhyme Book* of the improbable names of plants and animals, ending,

> No dandelions tell the time,
> Although they turn to clocks;
> Cat's cradle does not hold the cat,
> Nor foxgloves fit the fox.

Some species of digitalis are the source of the modern cardiac stimulant digitalin, a powerful glycoside that is used in the treatment of heart disease. But English apothecary Nicholas Culpeper said in the seventeenth century that it was used as a cure for "scabby heads" and as a remedy for king's evil or scrofula, an inflammation of the lymph glands, which from the time of Edward the Confessor could supposedly be cured only by the king or queen.

The medicinal properties of digitalis have been known for centuries. However, an old Welsh legend claims to be the first to prescribe it, because the knowledge of its properties came to the *meddygon*, the Welsh physicians, in a magical way. The legend is loosely based on the early thirteenth-century historical figure Rhiwallon, the physician to Prince Rhys the Hoarse, of South Wales. Young Rhiwallon was walking beside a lake one evening when from the mist arose a golden boat. A beautiful maiden was rowing the boat with golden oars. She glided softly away in the mist before he could speak to her. Rhiwallon returned every evening looking for the maiden; when he did not find her, he asked advice from a wise man. He told Rhiwallon to offer her cheese. Rhiwallon did as he was told, and the maiden appeared and took his offering. She came ashore, became his wife, and bore him three sons. After the sons grew and the youngest became a young man, Rhiwallon's wife rowed into the lake one day and returned with a magic box hinged with jewels. She told Rhiwallon that he must strike her three times so that she could return to the mist forever. He refused to hit her, but the next morning as he finished breakfast and prepared to go to the fields to work, Rhiwallon tapped his wife affectionately on the shoulder three times. Instantly a cloud of mist enveloped her and she disappeared. Left behind was the bejeweled magic box. When the three sons opened it, they found a list of all the medicinal herbs, including the foxglove, with full directions for their use and their healing properties. With this knowledge the sons became the most famous of physicians.

In the second part of the book-length poem *The Botanic Garden* (1791), Erasmus Darwin described Hygeia, the Greek goddess of health or hygiene, in the garb of the digitalis:

> Divine Hygeia, from the bending sky
> Descending, . . .
> Assumes bright Digitalis' dress and air,
> Her ruby cheek, white neck, and raven hair.
> Four youths protect her from the circling throng,
> And like the Nymph the Goddess steps along.

Darwin added in a footnote that foxglove was used to cure dropsy "where the legs and thighs are much swelled" in those "people past the meridian of their life." This cure was applied in *Silas Marner* (1861), George Eliot's story of a linen weaver, when Silas treats a cobbler's wife suffering from dropsy, "Recalling the relief his mother had found from a simple preparation of foxglove."

The foxglove has accumulated several symbolic meanings. Sir Walter Scott's *Lady of the Lake*, a poem in six cantos published in 1810, includes a note on foxglove: "Fox-glove and nightshade, side by side,/ Emblems of punishment and pride." And in the language of flowers, foxglove suggested both insincerity and "a wish."

For centuries poets have been fascinated by the bell-shaped foxglove and have often included it in their floral descriptions. Christina Rossetti described the tall foxglove in "A Bride Song":

> Through the vales to my love!
> Where the turf is so soft to the feet
> And the thyme makes it sweet,
> And the stately foxglove
> Hangs silent its exquisite bells.

In the late 1800s Thomas Edward Brown wrote "White Foxglove" in which he addressed the flower, but it shrank away and did not respond:

> White foxglove, by an angle in the wall,
> Secluded, tall,
> No vulgar bees
> Consult you, wondering
> If such a dainty thing
> Can give them ease.

In *Poetical Works* (1821), a collection by John Keats, a sonnet describes a place "Where the deer's swift leap/Startles the wild bee from the foxglove bell." In Coleridge's "The Keepsake" (1817), the passing of summer is marked when

> the foxglove tall
> Sheds its loose purple bells, or in the gust,
> Or when it bends beneath the upspringing lark,
> Or mountain-finch alighting.

In book 8 of *The Prelude*, Wordsworth, reflecting on his youthful poetic faculties, wrote of the same season:

> Through quaint obliquities I might pursue
> These cravings; when the foxglove, one by one
> Upwards through every stage of the tall stem
> Had shed beside the public way its bells,
> And stood of all dismantled, save the last
> Left at the tappering ladder's top, that seemed
> To bend as doth a slender blade of grass.

Many poets have been struck by the beauty of foxgloves. Shakespeare overlooked the distinctive flower, probably because it is not common in the Midlands. But Jean Ingelow, Victorian poet and children's author, wrote in "Divided" (1863) of foxgloves as an integral part of the natural landscape:

> An empty sky, a world of heather,
> Purple of foxglove, yellow of broom;
> We two among them wading together,
> Shaking out honey, treading perfume.

In his elegy "In Memoriam" (1833, published in 1850) to friend Arthur Henry Hallam, who died suddenly at the age of twenty-two of a cerebral hemorrhage, Tennyson expressed a profound sense of loss and regret for his friend's death and looked expectantly to the return of the new year to "burst the frozen bud":

> Bring orchis, bring the foxglove spire,
> The little speedwell's darling blue,

> Deep tulips dash'd with fiery dew,
> Laburnums, dropping-wells of fire.

Through all the long history of the plant, tall foxgloves have graced the back borders of gardens. On a quiet spring evening one can almost hear the sound of little bells tinkling at one's feet.

FRITILLARY

A diadem of gold, and richly grac'd

Fritillaries, with bell-shaped flowers, are members of the Liliaceae. The 100 bulbous, perennial species of *Fritillaria* are native to the Northern Hemisphere and are found in the Mediterranean region, Asia, and western North America.

The genus name *Fritillaria* is derived from the Latin *fritillius*, meaning dice box, an allusion to the doubled, truncated cone, almost box-like, shape of the flower, such as in the common European species, *F. meleagris*. The common name of this species is guinea flower, because its spotted markings resemble those of the guinea hen; *meleagris* is the Latin word for that fowl. An elegant species native to Turkey, *F. imperialis* is called the crown imperial, because of its vague resemblance to a monarch's crown and because it was first grown in the Imperial Gardens in Vienna; its bulbs have a musky scent strongly resembling that of a fox. A more diminutive species is *F. michailovskyi*, whose flowers are a curious mauve-brown and yellow. In some herbals the fritillary was called a tulip or a checkered daffodil.

According to a central European legend, the crown imperial grew in the Garden of Gethsemane and at one time had a white bloom. During the agony and prayers of Jesus, all flowers bowed their heads to the earth except the crown imperial; it alone remained upright. As Jesus was led away he looked at the flower that was not bowing, and contrition filled its heart. Ever since then, the crown imperial has humbly bent its head, and its white petals, now reddish, have borne the blush of shame.

A story from ancient Persia tells of a queen who was so beautiful that she made the king jealous. In a rage he unjustly banished her from the palace. But a compassionate angel rooted the queen's feet in the soil and changed her into the crown imperial so that her dignity and beauty would still command the respect and affection they had before her transformation.

The theme of fritillary as royalty has been used in numerous poetic works. An unknown author penned these beautiful lines in "The Imperial Flower":

> This lily's height bespeaks command,
> A fair, imperial flower;
> She seems designed for Flora's hand,
> The scepter of her power.

René Rapin expanded the royal metaphor:

> Then her gay gilded Front th' Imperial-Crown
> Erects aloft, and with a scornful Frown
> O'erlooks the subject Plants, while humbly they
> Wait round, and Homage to her Highness pay;
> Beneath the Summit of her Stem is plac'd
> A Diadem of Gold, and richly grac'd.
> Then verdant Leaves in bushy Plumes arise,
> And crisp'd and curling entertain the Eyes;
> Beneath these Leaves four radiant Blossoms bent
> Like painted Cups revers'd are downward sent
> No Flow'rs aspire in Pomp and State more high,
> Or lays a juster Claim to Majesty.

Conversely, George Herbert, a deeply religious poet of the early seventeenth century, in "Peace" (1633) used the crown imperial to show the decay of royal rule:

Then went I to a garden and did spy a gallant flower,
The Crown Imperial: sure, said I,
Peace at the root must dwell;
But when I digged I saw a worm devour
What shewed so well.

Mary Botham Howitt wrote in "The Wild Fritillary,"

Like a drooping thing of sorrow,
Sad to-day, more sad to-morrow;
Like a widow dark weeds wearing,
Anguish in her bosom bearing.

. . . .

Like a joy my memory knoweth—
In my native fields it groweth;
Like the voice of one long parted,
Calling to the faithful-hearted;

Like an expected pleasure
That hath neither stint nor measure;
Like a bountiful good fairy,
Do I hail, Fritillary.

John Langhorne wrote of the fritillary in *Fables of Flora* in a poem titled "The Queen of the Meadow and the Crown Imperial." He described it as a flower of "haughty mien":

In all the pomp of eastern state,
In all the eastern glory gay,
He bade with nature pride elate
Each flower of humbler birth obey.

And the fritillary speaks,

"In climes of orient glory born,
Where beauty first and empire grew;

Where first unfolds the golden morn,
Where richer falls the fragrant dew.

Then lowly bow, ye British flowers!
Confer your monarch's mighty sway,
And own the only glory yours,
When fear flies trembling to obey."

GENTIAN

Blue—blue—as if that sky let fall

Found naturally worldwide except in Africa, the 400 species of *Gentiana* are primarily perennials—but some are annuals and biennials—in the Gentianaceae. They generally bloom in autumn. Although the most well-known species have petals that glow true blue, some have funnel-shaped flowers with four to seven lobes and come in yellow, red, or white. *Gentiana acaulis* comes from the mountains of Europe and grows up to 10 centimeters (4 inches) tall, making it a valuable gentian of rock gardens; the bright yellow *G. lutea* grows up to one and a half meters (5 feet) tall. This tall plant is the commercial source of gentian root tonic, which is used to flavor liquors, such as bitters, and to aid digestion in after-dinner cordials.

The closed or bottle gentian, *G. andrewsii*, of eastern North America provides a stunning sight in blue and white along woodland borders in the autumn. *Gentiana verna*, as its name states, heralds the coming of spring in mountain meadows throughout the European Alps. *Gentiana autumnalis*, the pine barrens gentian of the eastern United States, grows to about a meter (3 feet) tall; it has a splendid related form, *G. autumnalis* f. *albescens*.

The name of the genus is derived from King Gentius, the second-century ruler of Illyria, an ancient country that lay east of the Adriatic. Gentius is credited with discovering the medicinal virtues of the root of the yellow gentian, having used it to cure a mysterious fever which afflicted his army. Old names for this species are bitterwort and felwort, the latter because the gentian was thought to cure felons, abscesses near the fingertips, and galls, sores caused by chafing. Pliny the Elder called the marsh gentian, *Gentiana pneumonanthe*, the "Calathian violet—the gift of autumn." The species *G. andrewsii* is named after a nineteenth-century English flower painter, Henry C. Andrews; *G. pneumonanthe*, or lung-flower, was probably used to treat pulmonary problems; *G. acaulis* means "without a stem"; *G. lutea* means yellow; and *G. autumnalis* means autumn bloomer. In the Victorian language of flowers, gentians said, "You are unjust."

Native North Americans used a brew of gentian root to soothe stomach aches and to purify the blood. American settlers, drawing upon European folk medicine, used a piece of gentian root in brandy, whiskey, or gin to stimulate the appetite or, having eaten a meal, to aid in digestion.

In a fifteenth-century Christian Hungarian legend, King Ladislaus asked God for help to overcome a plague that was afflicting his people, who were dying by the thousands. He prayed that if he shot an arrow into the air it would be guided to a plant that would provide the cure and check the pestilence. His arrow pierced the ground, resting on a gentian root. The root was immediately used for treatment, and the ailing were astonishingly cured.

The virtues of gentians were described by John Gerard (1633):

> The root of gentian given in pouder the quantitie of a dramme, with a little pepper and herbe Grace mixed therewith, is profitable for them that are bitten or stung with any mannor of venomous beast or mad dog; or for any that hath taken poison. The decoction drunke is good against the stoppings of the liver, and cruditie of the stomacke, helpeth digestion, dissolueth and scattereth congealed bloud, and is good against all cold diseases of the inward parts.

Of the bastard felwort—possibly the spring gentian, *Gentiana verna*—Gerard says, "At the top [of the plant] stands a cup, out of where comes one longe floure without smell, and as it were divided at the top into five parts; and it is of so elegant a colour, that it seems to exceed blewnesse it selfe."

The Cornhill Magazine (1892), an English literary monthly published on London's Cornhill Street and first edited by William Thackeray, described the use of gentian root in spirits: "The gentian spirit may be said to be the very elixir of life to the mountain folk . . . . The smell of gentian brandy is not pleasant, especially if new, but with age the spirit greatly improves, mellows, and loses its disagreeable aroma."

A fairy tale of the fringed gentian (formerly *Gentiana crinita*, but now a related genus *Gentianopsis crinita*) tells the story of the queen of fairies who stayed out late on Halloween night. After midnight, the friendly fireflies went to bed, leaving her alone and frightened. She asked a gentian if she might sleep in its blossom until daybreak, but the gentian was sleepy and did not want to be disturbed. "Who are you anyway?" he grumbled. When the fairy queen told him who she was, the gentian said, "Well, then you ought to be able to find some other flower." But it was late October and only the gentians were blooming. She searched more, and finally one lone gentian invited her in to spend the night. At sunrise, she left, kissing the little gentian that had offered her shelter. The fairy queen said, "I shall make you different from all the other gentians." And she gave the fringed gentian the power to open during the day to feel the sun and to close at night to bathe in the dew.

In keeping with the special place of the fringed gentian, many poets set it apart from other gentians. William Cullen Bryant wrote the eloquent paean "To the Fringed Gentian" (1856):

> Thou waitest late, and com'st alone,
> When woods are bare and birds are flown,
> And frosts and shortening days portend
> The aged Year is near his end.

Helen Hunt Jackson, a poet of the mid-1800s, used it as a sign of the season in "October's Bright Blue Weather":

> When Gentians roll their fringes tight,
> To save them for the morning,
> And chestnuts fall from satin burrs,
> Without a sound of warning.

Sarah H. Whitman, American writer, poet, and close friend of Edgar Allan Poe, recalled the fringed gentian's dewy slumbers in "Closed, Sweet Dreams, I Look to Heaven":

> Beside the brook, on the umbered meadow,
> Where yellow fern tufts fleck the faded ground,
> With folded lids beneath the palmy shadow
> The gentian nods in dewy slumbers bound.

And Emily Dickinson in "Fringed Gentian" used the unique quality of the plant's fruition as a metaphor for individuality:

> God made a little gentian;
> It tried to be a rose
> And failed, and all the summer laughed.
> But just before the snows
> There came a purple creature
> That ravished all the hill;
> And summer hid her forehead,
> And mockery was still.
> The frosts were her condition;
> The Tyrian would not come
> Until the North evoked it.
> "Creator! shall I bloom?"

Most gentians may be blue in color as well as in the origin of their name, but they are finicky—perhaps like some ancient kings. They are

difficult to grow and uncommon, but once seen in woodlands or rock-
eries on a brilliant autumn day, they are a sight to be remembered.
Bryant best phrased such a moment of wonderment in his "To the
Fringed Gentian":

> Thou blossom, bright with autumn dew,
> And colored with the heaven's own blue,
> That openest when the quiet light
> Succeeds the keen and frosty night
>
> . . . .
>
> Then doth thy sweet and quiet eye
> Look through its fringes to the sky,
> Blue—blue—as if that sky let fall
> A flower from its cerulean wall.

GERANIUM & PELARGONIUM
Constantly courting my attention

Perennial herbs and small shrubs of the Geraniaceae, *Geranium*, or cranesbills, consist of about 300 species found worldwide in temperate and tropical regions. With numerous cultivars, these ornamentals have been garden features for years. Herb Robert, *G. robertianum*, for example, has been grown in gardens since at least the ninth century. The large-rooted geranium of Europe is *G. macrorrhizum*, and the native American geraniums include *G. maculatum* and *G. carolinianum*. One handsome meadow cranesbill is *G. pratense*, which has showy pale blue flowers tinged with pink. Its flowers were formerly used for making a blue-gray dye, a sort of military blue. Well known is *G. sanguineum*, the bloody cranesbill, with striking magenta flowers.

The true geraniums, sometimes called English geraniums in old literature, are different from the greenhouse or bright window-box "geraniums" of primarily South African origin. These are storksbills of the genus *Pelargonium* and include various so-called scented-leaf or fancy geraniums. Unlike true geraniums, pelargoniums have irregular flower shapes. The *Erodium* or heronsbill is another relative. All three genera are closely related botanically within the Geraniaceae. They are generally distinguished from one another by the number of fertile stamens.

Cranesbill, storksbill, and heronsbill are contemporary common names in wide use, but the roots of these names travel back to Greek and other mythologies, suggesting a high degree of knowledge about ornithology and botany on the part of the creators of those myths. The name *Geranium*—more often used as a common name for pelargoniums in the United States—derives from *geranion*, the Greek name for the plant, which itself comes from *geranos*, for crane, an allusion to the shape of its fruit. Cranesbill is the common name most widely used for true geraniums in Europe. The name *Pelargonium* comes from the Greek word for storksbill. The confusion of the two plants traces to the 1700s when storksbills first arrived in Europe and were erroneously classified as geraniums. An interesting footnote to the origin of the geranium's name is that *geranos* in French became *grus* and finally *grue* for crane. That word is embedded in our English word *pedigree* from the French *pie de grue*, which translates as "the crane's foot." A pedigree chart is indeed reminiscent of a crane's footprint.

Herb Robert, with its pink-veined petals, is supposed to have gotten its name from one or more saints named Robert. It is recorded as Sancti Ruperti Herba by early writers. Saint Rupert supposedly cured various ailments using *Geranium robertianum*, and St. Robert, an eleventh-century abbot, used it to aid in healing wounds. The Germans used it to cure Ruprechts-Plage, a plague supposedly carried to central Europe by Robert, a duke of Normandy. Herb Robert is also called poor robin, which is a name Wordsworth used.

An Islamic legend shows geraniums to be a gift from Allah. The prophet Mahomet, or Mohammed, went bathing one day, leaving his shirt on the bank near a clump of drab mallow plants. When he returned from his swim, Mahomet found that the sacred garment had transformed the humble mallow into bright-blossomed geraniums. In Greek mythology, a flock of cranes warned Megarus that his father,

Zeus, was upset with all the evils of humans and that he planned to send a flood to cleanse the earth. The young Megarus heeded the warning and hastened to Mt. Gerania, where he waited out the flood until the waters receded.

In the language of flowers, the wild geranium meant steadfast piety, the rose or pink geranium signified preference, and the oak-leaved geranium, probably a pelargonium, suggested true friendship.

Pelargoniums being called geraniums began appearing widely in literature in the nineteenth century. Louisa May Alcott in *Little Women* (1868) described the "green leaves, and the scarlet flowers of Amy's pet geranium." E. M. Forster noted in *A Room with a View* (1908) that the villa Albert's "tortured garden was bright with geraniums and lobelias and polished shells." In *The Scarlet Pimpernel* (1905) by Baroness Orczy, the coffee room of the Fisherman's Rest, an inn on the Channel coast of England, is the site of much important action in the novel. The oak-beamed room, already blackened with age, is still inviting to the customers: "In the leaded window, high up, a row of pots of scarlet geraniums and blue larkspur gave the bright note of colour against the dull background of the oak." In Oscar Wilde's *Birthday of the Infanta* (1891), anthropomorphic flowers react to the invasion of their garden by an ugly dwarf. The tulips, lilies, and roses all cry out in disgust and, "Even the red geraniums, who did not usually give themselves airs, and were known to have a great many poor relations themselves, curled up in disgust when they saw him." And geraniums appear in Edgar Allan Poe's "Landor's Cottage" (1850), a nonfiction piece describing a valley he has come into on a walk near the Hudson River:

> The expanse of the green turf was relieved, here and there, by an occasional showy shrub, such as the hydrangea, or the common snow-ball, or the aromatic seringa [syringa]; or, more frequently, by a clump of geraniums blossoming gorgeously in great varieties. These latter grew in pots which were carefully buried in the soil, so as to give the plants the

appearance of being indigenous. Besides all this, the lawn's velvet was exquisitely spotted with sheep—a considerable flock of which roamed about the vale, in company with three tamed deer and a vast number of brilliantly-plumed ducks.

In "Evelyn Hope" (1855) from Robert Browning, the narrator laments the death of sixteen-year-old Evelyn. Though he was "thrice as old" and she had scarcely heard his name, in his heart he had a place to spare for her smile:

> Beautiful Evelyn Hope is dead!
> Sit and watch by her side an hour.
> That is her book-shelf, this her bed;
> She plucked that piece of geranium-flower,
> Beginning to die too, in the glass;
> Little has yet been changed, I think:
> The shutters are shut, no light may pass
> Save two long rays through the hinge's chink.
>
> . . . .
>
> But the time will come—at last it will—
> When, Evelyn Hope, what meant, I shall say,
> In the lower earth,—in the years long still,—
> That body and soul so gay?
> Why your hair was amber I shall divine,
> And your mouth of your own geranium's red,—
> And what you would do with me, in fine,
> In the new life come in the old one's stead.

Wordsworth focused attention directly on the true geranium in "Poor Robin":

> Now when the primrose makes a splendid show,
> And lilies face the March winds in full blow,
> And humbler growths as moved with one desire
> Put on, to welcome spring, their best attire,

> Poor Robin is yet flowerless; but how gay
> With his red stalks upon this sunny day!
> And, as his tufts of leaves he spreads, content
> With a hard bed and scanty nourishment
> Mixed with the green, some shine not lacking power
> To rival summer's brightest scarlet flower;
> And flowers they well might seem to passers-by
> If looked at only with a careless eye;
> Flowers—or a richer produce (did it suit
> The season) sprinklings of ripe strawberry fruit.

Wordsworth appended to the poem "This little wild flower—Poor Robin—is here constantly courting my attention, and exciting what may be called a domestic interest with the varying aspects of its stalks and leaves and flowers. Strangely do the tastes of men differ according to their employment and habits of life. 'What a nice wall would that be,' said a labouring man to me one day, 'if all that rubbish was cleared off.'"

American poet Clement Wood wrote "Rose-Geranium" early in his career:

> A pungent spray of rose-geranium—
> A breath of the old life.
>
> . . . .
>
> And little grandson smell this spray of rose-geranium—
> Just think, when grandmother was a little tiny girl
> Her grandmother grew them in her yard!

And William Cowper noted that "Geranium boasts/Her crimson honours" in *The Task* (1785).

The North American species of *Geranium* were observed by the American naturalist and essayist Henry David Thoreau. Recorded in his journal on 19 June 1852 are geraniums native to New England: "Buttercups and geraniums cover the meadows, the latter appearing to float on the grass. It has lasted long, this rather tender flower."

GLADIOLUS

The sword-lily that cuts the clear waters

Gladiolus is a genus of old world plants from Europe, the Mediterranean region, and southern Africa. These herbaceous perennials in the Iridaceae grow from rhizome-like corms. The large-flowering hybrids generally dominate in gardens, though they were derived from only about seven species of the 180 that are known. *Gladiolus* ×*hortulanus* is the florist's gladiolus, a name given to a group of hybrids involving various species, and of which there are hundreds of cultivars. Because of the complexity of these hybrids, most are classified by flower color, size, shape, and form. *Gladiolus cardinalis* (the waterfall glad), *G. tristis*, and *G. dalenii* are three South African species that have contributed their genes to the garden cultivars. *Gladiolus illyricus* is native to southern England as well as the Mediterranean.

In ancient Rome a sword was called a *gladius* and a small sword was a *gladiolus*. Linnaeus, perhaps borrowing from Pliny's writings about plants with sword-shaped leaves, named the plant *gladiolus* in allusion to the shape of the narrow leaves. The word is also the root of the word gladiator, one who carried swords into the coliseums and who "lived or died by the sword." Likely drawing from the reputation of its namesake, the gladiolus suggested strength of character in the language of flowers. Glads have been called sword-lilies because of their Latin name and corn-flag because they are pests in corn or grain fields. Of the various specific epithets, *hortulanus* means of gardens, *cardinalis* means red or scarlet, and *tristis* means sad. The species *illyricus* means of Illyria, an ancient realm near the Adriatic Sea.

The gladiolus figured in Greek mythology. Some writers suggest that in Ovid's telling of the story of the death of Hyacinthus, who was killed by a discus errantly thrown by Apollo, he meant the gladiolus, not the hyacinth, sprang from the slain body. In another myth, Ceres, the Roman goddess of grain and harvest, Demeter in Greek mythology, loved a sacred grove near Thessaly. An evil and wealthy man named Erisichthon, who did not believe in the gods, lived nearby and freely took firewood from the sacred grove of trees. On one occasion when the worshippers tried to stop him, he cut off a man's head. From the blood, Ceres made to spring up little sword-shaped plants which she called gladiolus. Quick with revenge, Ceres punished Erisichthon by ordering Famine to enter his body. Unable to get enough food to satisfy his appetite, he sold his daughter in order to buy more food. The daughter escaped to the forest, and Ceres turned her into a gladiolus plant to watch over the man slain by her father. When Erisichthon could get no more food and had no more money, his craving for food was so strong that Famine forced him to eat himself.

Among the plant's earliest appearances in English writing is John Maplet's *A Greene Forest, or a Naturalle Historie* (1567) in which he wrote, "Gladiolus, his form and proportion of leafe is like to Sedge, his flower yealow in a maner like to the flower De luce." In literature, the gladiolus is sometimes called the corn-flag or sword-lily. Ralph Waldo Emerson wrote in his *Essays* (1841–1844), "You shall still see . . . the tasselled grass, or the corn-flags." And Robert Browning in "The Flight of Duchess" (1845) used the plant to paint a vivid scene: "Where the bold sword-lily cuts the clear waters." However, he might have been referring to *Iris pseudacorus*, which tolerates waterlogging.

The corn-flag or sword-lily was described and illustrated in John Gerard's great work of the seventeenth century, *The Herbal or General History of Plants* (1633 revision by Thomas Johnson). He recognized the French, Italian, and water corn-flags and noted that they grow in meadows and cornfields. Among the gladiolus's virtues were medicinal uses: "The roote stamped with the powder of Frankincense and wine, and applied, draweth forth splinters and thornes that sticke fast in the flesh. Being stamped with the meal of Darnell and honied water, doth waste and make subtill hard lumps, nodes, and swellings, being emplaistred." Finally, Gerard warned, "Some affirme, that the upper root provoketh bodily lust, and the lower [root] causeth barrennesse" because the old wrinkled corm remain beneath the new.

Writing at about the same time as Gerard, John Parkinson in 1629 described the gladiolus and additional forms of the corn-flag: the Constantinople, the blue, the white, and the small purple corn-flag, as well as the Italian and French that Gerard noted. Regarding the plant's virtues, Parkinson said that he doubted its ability to cure "venerie," venereal disease, but he noted the physician Galen, a Greek of the second century who authored about 500 works, "giveth unto it a drawing, digesting, and drying faculty."

GUERNSEY LILY

The radiant tinctures may with tap'stry vye

The Guernsey lily or Guernsey flower is *Nerine sarniensis*, an amaryllis-like plant. In all, the genus *Nerine* has thirty species with strap-shaped leaves and handsome white, pink, or red flowers borne on tall scapes. All the species are native to southern Africa and are frost tender in temperate regions.

The name *Nerine* is from Nereis, one of the fifty beautiful maiden sea nymphs called the nereids, who were daughters of Nereus and Doris of Greek mythology. The nereids were helpful to voyagers and mariners in distress. They are depicted in art either lightly draped or nude in poses that harmonize with the ocean waves of their element. The specific epithet *sarniensis* means of Guernsey, one of the Channel Islands, the ancient name for which is Sarnia.

How the Guernsey lily obtained its common name is at once confusing and intriguing. It has been widely accepted since first reported in the 1680s by Oxford botany professor Robert Morison, an English Royalist, that the lily arrived and naturalized in the Channel Islands after the wreck off the Guernsey coast of a ship bearing the plants from Japan. However, the plant is known to have been growing in the Parisian garden of Jean Morin in 1634 and in the Wimbledon garden of Cromwellian supporter John Lambert in 1650. In reality, political ruse appeared to be at play in Morison's story. Lambert was exiled to Guernsey from 1661 to 1670 following the monarchy's restoration; he took with him his plants, which had been obtained through an uncertain route from South Africa, not Japan. Whether or not the shipwreck story is true, the sand dunes of Guernsey are probably the only place in northern Europe in which *Nerine sarniensis* could survive, for it needs to grow in sand and a frost-free climate.

The association of *Nerine sarniensis* with Japan has been embedded in numerous writings. For example, John Evelyn in his *Kalendarium Hortensis* (1664) called it "the narcissus of Japan or Guernsey Lilly." As recently as 1838 the authors of the *Penny Cyclopedia of the Society for the Diffusion of Useful Knowledge* concluded, "The Guernsey lily, a species of the amaryllis, is a native of Japan."

A tale of how the Guernsey lily first appeared on the island of Guernsey tells that the fairy king won the hand and heart of the beautiful Michelle and tried to take her away to his kingdom. However, Michelle did not want to leave her family, and she asked the king to give her a token by which her family could remember her. He gave her the bulb of a lily, which she planted and which burst into flower after she left Guernsey. When her mother went looking for Michelle, she found the bulb in full bloom and knew immediately that her daughter had planted it.

The Lusiad by Luis de Camoëns, a Portuguese poet of the sixteenth century, is the national epic poem of Portugal that tells of the glorious achievements of Portuguese explorers who came after Lusus, the legendary founder of Lusitania, or Portugal. The explorers encounter Bacchus, who plans to destroy the Portuguese at Mombasa; but Venus, with the aid of the nereids, the sea-guardians of the virtuous, staves him off. Jupiter watches the ride of the nereids but does not intercede on either side. In the following selection, translated by Sir Richard Fanshawe in 1655, Nerine, whose name was chosen by William Herbert for the Guernsey lily later in the nineteenth century, rides the waves with Venus and other sea nymphs:

> Now through the ocean in great haste they flunder,
> Raising the white foam with their silver tayles.
> Cloto with bosom breaks the waves in sunder,
> And, with more fury then of custom, sayles;
> Nise runs up an end, Nerine (younger)
> Leaps o'er them; frizled with her touching Scales,

The crooked Billows (yielding) make a lane
For the feard Nymphs to post if through the Maine.

And René Rapin, in *De Hortorum Cultura*, described a "purple nar-cissus" that most assume to be *Nerine sarniensis*:

Late from Japan's remotest Regions sent,
Narcissus came array'd in purple Paint,
And numerous Spots of yellow stain the Flower,
As richly sprinkled with a golden shower:
The radiant Tinctures may with tap'stry vye,
And proudly emulate the Tyrian dye.

HEATH & HEATHER

Wild blossoms of the moorland

Heaths and heathers belong to the Ericaceae, a vast family of 116 genera and more than 3000 species of shrubs, trees, herbs, and even saprophytes, scavenger plants that lack chlorophyll and find nourishment from dead or decayed organic matter. Numerous species of the family are grown as ornamentals. The heaths and heathers, constituting about one-fourth of all species in the family, are members of the genus *Erica*, an evergreen whose native distribution is Europe, the Middle East, and Africa. Well-known garden species, some with numerous cultivars, include *Erica carnea*, winter heath; *E. cinerea*, bell heather; *E. erigena*, Irish heath; and *E. vagans*, Cornish heath. Closely related to erica is the monospecific genus *Calluna*, the Scottish heather or ling, consisting of *C. vulgaris* and its hundreds of cultivars. It is native to European moors and other uncultivated lands; it has become naturalized in parts of North America where its pink-lobed flowers bloom in the late summer and autumn.

The name erica is derived from the Greek *ereike*, meaning "to break or crush," in the belief that an infusion of heather tea would break up bladder stones. The common names descend from Old English: heath from *haeth*, meaning an untilled track of land, and heather from *heddre*, a place name, possibly for a particular heath. It is also the root of the word heathen, someone living away from the church in the "wilderness." The latter seems to have been influenced by Scottish Gaelic since it appeared in that language as early as the 1300s in the similar word *hathir*. *Erica* is also the classical name for the tree heath, *E. arborea*, which is woody and grows to 4 meters (13 feet), whose burl (root knot) was used to make tobacco pipes. Such pipes have been called briar or brier pipes from the old Middle French words for heath and heather, *bruyere* or *bruiere*.

The name of *Calluna* is derived from the Latin word *kalluno*, meaning to clean, because the heather was used to make brooms. It was also used as thatching for houses, yellow dye, mattress bedstraw, and for the making of a folk brew known as heather ale. The Picts, a non-Celtic people who had amalgamated with the Scottish by the ninth century, made beer from heather flowers in a secret method that was not to be shared with outsiders. By the nineteenth century, a "potable liquor" was being brewed in some remote districts of Scotland by mixing heather and hops.

In literature, references to heaths and heathers and their flowers appear in a variety of sources. For example, in Shakespeare's *The Tempest* (1611), Gonzalo, "an honest old counsellor," is about to be shipwrecked in a storm and he shouts, "Now would I give a thousand furlongs of sea for an acre of barren ground; long heath, brown furze, any thing. The will above be done! but I would fain die a dry death."

The Bible makes reference to heath twice in the book of Jeremiah, though the plants are unlikely to be species of *Erica*. In chapter seventeen, verse six, the Lord describes the punishment for idolatry and the judgment for the man "whose heart departeth from the Lord":

> For he shall be like the heath in the desert,
> and shall not see when good cometh; but
> shall inhabit the parched places in the
> wilderness, in a salt land and not
> inhabited.

And later, in chapter forty-eight, verse six, the kingdom of Moab, part of what is now Jordan, is judged by the Lord for its arrogance:

Flee, save your lives, and be like the heath in the
wilderness.

A traditional use of heather was as bedstraw, usually arranged so
that the flowers were at the head of the bed. Sir Walter Scott described
this practice in *The Lady of the Lake* (1810), which is set in Loch Katrine,
Scotland:

> The hall was clear'd; the stranger's bed
> Was there of mountain heather spread,
> Where oft a hundred guests had lain,
> And dream'd their forest sports again.
> But vainly did the heath flower shed
> Its moorland fragrance round his head.

Scott again mentioned the use of heather for bedding in *Rob Roy* (1817),
which is set in England during the Jacobite uprising of the early 1700s:

> I remarked that Rob Roy's attention had extended itself to
> providing us better bedding than we had the night before.
> Two of the least fragile of the bedsteads, which stood by
> the wall of the hut, had been stuffed with heath, then in full
> flower, so artificially arranged that the flowers, being upper-
> most, afforded a mattress at once elastic and fragrant.
> Cloaks, and such bedding as could be collected, stretched
> over this vegetable couch, made it both soft and warm.

James Hogg, dubbed the "Ettrick Forest Shepherd" for his place
of birth and line of work, also wrote of heather bedstraw—for a bird
who sings over the moorlands and the lea—in "The Skylark":

> Musical cherub, soar, singing, away!
> Then, when the gloaming comes,
> Low in the heather blooms
> Sweet will thy welcome and bed of love be!

Wordsworth, who has been both criticized as simple and dull and venerated as insightful of humanity and nature, used heather as a symbol of love in "Yarrow Visited" (1814). On a visit to the River Yarrow in Scotland he recalled his childhood and "the brood of chaste affection":

> How sweet, on this autumnal day,
> The wildwood fruits to gather
> And on my truelove's forehead plant
> A crest of blooming heather!
> And what if I enwreathed my own!
> 'T were no offence to reason;
> The sober hills thus deck their brows
> To meet the wintry season.

The longing for homeland and the sight of flowering heather is told with an aching heart in these verses by Anne MacVicar Grant of Laggan, Scotland, titled "To the Heather":

> Flowers of the wild! whose purple glow
> Adorns the dusky mountain's side,
> Not the gay hues of Iris' bow,
> Nor garden's artful, varied pride,
> With all its wealth of sweets could cheer,
> Like thee, the hardy mountaineer.
>
> Flowers of his dear-loved native land!
> Alas! when distant, far more dear.
> When I, from cold and foreign strand,
> Look homeward through the blinding tear,
> How must his aching heart deplore
> That home, and thee, I see no more!

In chapter 16 of *Wuthering Heights* (1847) by Emily Brontë, Catherine dies during childbirth and is buried on a moor covered with heath:

The place of Catherine's interment, to the surprise of the villagers, was neither in the chapel under the carved monument of the Lintons, nor yet by the tombs of her own relations, outside. It was dug on a green slope in a corner of the kirkyard, where the wall is so low that heath and bilberry plants have climbed over it from the moor; and peat mould almost buries it. Her husband lies in the same spot now; and they have each a simple headstone above, and a plain gray block at their feet, to mark the graves.

In the language of flowers, the heath plant represented solitude, no doubt reflecting the quietness and isolation of the moorlands on which it grows. Eliza Cook described that isolation and loneliness in "Moorland Blossoms":

Wild blossoms of the moorland, ye are very dear to me;
Ye lure my dreaming memory as clover does the bee;
Ye bring back all my childhood loved, when freedom, joy,
    and health
Had never thought of wearing chains to fetter fame and
    wealth.
Wild blossoms of the common land, brave tenants of the
    earth,
Your breathings were among the first that helped my spirit's
    birth;
For how my busy brain would dream, and how my heart
    would burn,
Where gorse and heather flung their arms above the forest
    fern.

Thomas Hardy mentioned heath and heather several times in *The Return of the Native* (1878). Eustacia Vye, who marries Clym Yeobright, stands on the beautiful Egdon Heath and listens to the November winds knife through "the mummied heathbells of the past summer, originally tender and purple, now washed colourless by Michaelmas rains and dried to the dead skins by October suns." But in the following summer the heather flowers gloriously: "The July sun shone over Egdon and fired its crimson heather to scarlet. It was the one season of the year, and the one weather of the season, in which heath was gorgeous." But the seasons are unrelenting: "The sun had branded the whole heath with his mark, even the purple heath flowers having put on a brownness under the dry blazes of the few preceding days."

In the opening lines of Tennyson's poem "Maud" (1854), the narrator, a man of morbid nature, describes the place of his father's death:

> I hate the dreadful hollow behind the little wood,
> Its lips in the field above are dabbled with blood-red heath,
> The red ribb'd ledges drip with a silent horror of blood,
> And Echo there, whatever is ask'd her, answers Death.

"The Excursion" (1814) by William Wordsworth is the companion to his great but unfinished poem "The Recluse" (1806). In "The Excursion," searchers discover an old man who was feared drowned sheltered among the heath during a storm:

> And long and hopelessly we sought in vain:
> Till, chancing on that lofty ridge to pass
> A heap of ruin—almost without walls
> And wholly without roof (the bleached remains
> Of a small chapel, where, in ancient time,
> The peasants of these lonely valleys used
> To meet for worship on that central height)—
> We there espied the object of our search,
> Lying full three parts buried among tufts
> Of heath-plant, under and above him strewn,
> To baffle, as he might, the watery storm.

Finally, the German poet Krummacher wrote of finding God's presence in "Alpine Heights" (translated by Charles T. Brooks), where the mountains are covered with "flowerets white and blue" and a glacier of ice that "gleams like a paradise":

> On Alpine heights, o'er many a fragrant heath,
> The loveliest breezes breathe;
> So free and pure the air,
> His breath seems floating there,
> On Alpine heights a lovely Father dwells.

HELLEBORE

Bright as the silvery plume

Although hellebores are botanically members of the buttercup family, Ranunculaceae, the flowers of the more familiar species have large rose-like sepals, considered petals. The fifteen species of this perennial have a native distribution from Europe to Asia. *Helleborus niger* is the Christmas rose; other well-known species include *H. ×hybridus* (a garden hybrid complex, formerly *Helleborus orientalis*), *H. argutifolius*, and *H. foetidus.*

The Latin name *Helleborus* is from the Greek *helleboros*, a combination of *helein*, to kill, and *bora*, food; the name appears to mean "food that kills." From at least the 1500s in carols of Advent, the season of the Christian church just before Christmas, and the 1600s in writings of herbalists, the name rose has been incorrectly applied to hellebores. The English names Christmas rose and Lenten rose are widely used for *Helleborus niger* and *H. ×hybridus*, respectively. In addition *H. niger* is known by its more poetic names Holy Night rose, the rose of Noel, and the Christmas bloom. In Dutch it is called Christ's herb. Because hellebores have large flowers and bloom so prominently in winter, they are often more noticeable than smaller plants of that season.

Various legends relate to *Helleborus niger's* Christmastide blooms. In one account, the flower originated through a poor young shepherdess, Madelon, who followed the other shepherds to Bethlehem to visit the baby Jesus. On the way, she wept because she had no gift to give him, but the angel Gabriel appeared and comforted her. He touched his staff to the cold, frozen ground, and brought forth a lovely bunch of hellebores so that Madelon could bear a gift. Some versions of this story claim a rose sprang from the ground that was touched by her tears. She picked the flowers and laid them in the manger; and when the child touched the petals with his rosy fingers, they turned pink.

Another story was written by Swedish author Selma Lagerlöf in the December 1907 issue of *Good Housekeeping* magazine. In 1909 Lagerlöf became the first woman to win a Nobel Prize for Literature for her writings of European legends and folklore. According to her story, long ago, every Christmas Eve a portion of Göinge Forest in Sweden turned into a beautiful garden. The sole witnesses of the transformation, however, were the poor Robber family, who had been banished from the city for stealing a neighbor's cow for food. Hoping to get her husband pardoned, the exiled wife agreed to reveal the location of the flower garden to the abbot and a gardener brother one Christmas Eve. The brother, jealous because his own garden in the abbey was not nearly so beautiful as that in the forest, cursed the garden, saying that it must be the work of the Evil One if the garden was known only to outlaws. At his words, darkness sank over the garden and it became frozen and covered with snow. The old abbot died in the sudden snow, clutching a pair of roots in his hand. The Christmas garden never bloomed again in Göinge Forest, but the abbot's roots were planted at the abbey, and now, every Christmas season, the plant sends forth green stalks with white petals and golden stamens. The brothers call it the Christmas rose.

A great carol of the Christmas season, most commonly sung by Europeans, uses the Christmas rose as a symbol for the Christ child. In 1529, a prayer book with music was printed, containing a folk carol of nineteen stanzas that opens with the German words *Es ist ein Reis entsprungen* (There a branch has sprouted). The carol appears to date from even a century earlier. Some versions substitute *Ros'* (rose) for *Reis* (branch). In English it is the familiar carol "Lo! How a Rose, E'er Blooming." One German version dating from the 1800s contains an astonishing twenty-two verses and is entitled *Winterblümlein*, "The Little Flower of Winter." Both the original German and various English translations of the carol speak of the blooming of a rose on a cold, snowy night at Christmastime. The rose cannot be a *Rosa* but is most likely a *Helleborus niger*. As a translation of the carol from the late 1800s explains:

> From noble root new sprung,
> With thornless branch extended
> A rose doth bear a flower.
> All in the cold midwinter
> And at the midnight hour.

The carol in its extended versions narrates the Christmas story from the Annunciation to conception to visits by the shepherds and the magi. Of particular interest is the mention of Jesse. Most English versions and the standard German version of 1609 by Praetorius include the words, "Of Jesse's line descended/By ancient sibyls sung." Jesse is the presumed father of King David, and according to messianic prophecy he was "the patriarch of Christ's genealogy." The prophecy declares that a "branch shall grow from his roots . . . and will bear a little flower." In medieval iconography, the tree of Jesse was frequently depicted as a rose plant. Thus, the flowering of the Christmas rose each season recalls the "little flower" of Jesse.

The hellebore has a long recorded history dating back to Hippocrates and Theophrastus, who nearly 2200 years ago knew it as a drug, purgative, and "poison paste to kill wolves and foxes." The plants were cultivated during the Middle Ages for medicinal purposes primarily in monastery gardens. In some cases it is well established that the Christmas rose is not the hellebore of these early writers. Nevertheless, the black hellebore, presumably *Helleborus niger*, was supposedly favored by witches who used it in their charms because they believed one "finger" of its lobed leaves was evil. According to legend, only a witch knows which one!

Pliny the Elder's great work, *Historia Naturalis*, consisting of some thirty-seven volumes, is a vast collection of both information and misinformation often derived from secondhand observations and superstitions. For example, the Gauls, when they "ride a hunting into the chase, used to dip their arrow heads in the juice of hellebore, and they have this opinion that the venison which they take will eat the tenderer." Pliny noted that the black hellebore root was used as a purgative and a cure for lunacy, but for horses, oxen, and swine, "it killeth them." His citations distinguish between the white and black hellebore (sometimes the white hellebore was a *Veratrum*). He listed twenty-four remedies derived from black hellebore and twenty-three from the white. The remedies relieve everything from itch in quadrupeds to "bilious secretions and morbid humours" in people and include uses as a rat and mouse poison. Pliny said that "it is universally recommended not to give hellebore to aged people or children, to persons of a soft and effeminate habit of body or mind, or of a delicate or tender constitution . . . [and] persons of a timorous disposition are recommended not to take it."

Traditionally, even the collecting of black hellebores was considered dangerous because of their connection to witchcraft and sorcery. It had to be done in a specific, prescribed way; Pliny instructed drawing a circle around the plant with a sword and while lifting the root saying certain spells or prayers, entreating permission from the gods. The mystic rites for collecting, according to some versions, suggest looking to the east to be sure that no eagle witnesses the process; if it does, the gatherer will waste away and die within a year.

Citations on hellebores date from as early as the 1000s, and many medieval herbalists used the plant. References to hellebores are found in the works of Sir Walter Scott and Alfred, Lord Tennyson, to name only two of the many. In Scott's *Letters on Demonology and Witchcraft* (1830), he wrote of "Wretches fitter for a course of Hellebore than for the stake." And Tennyson in "Becket" (1884) wrote, "Such strong hate-philtre as may madden him—madden Against his priest all hellebore." Lord Byron spoke of "tuns of helleboric juice" in "Hints from Horace" (1811), and Bishop Joseph Hall wrote of "errors that are more fit for hellebore than theological convictions" in *The Invisible World* (1652).

Chaucer listed hellebore as a purgative remedy for Chauntecleer, the cock in *The Nun's Priest's Tale* of *The Canterbury Tales* (1387–1400). According to the tale, one day Chauntecleer awakes from a nightmare, having dreamed that a fox is after him. Pertelote, his hen wife, says he has indigestion and recommends,

> A day or two ye shul have digestyves
> Of wormes, beffore you take your laxatyves
> Of spurge-laurel, centaury, and fumitory
> Or elles of [h]ellebore that groweth there,
> Of caper-spurge, or of dogwood's beryis.

Anne Brontë obviously knew her horticulture when Helen Graham, the heroine in Brontë's second and last novel, *The Tenant of Wildfell Hall* (1848), plucked a "half-blown Christmas rose," *Helleborus niger*, took it inside, dusted off the glittering powder of snow, kissed it, and handed it to her new lover Gilbert Markham, a young farmer. It was December, and the scene took place inside, near the bay windows at Wildfell Hall.

> This rose is not so fragrant as a summer flower, but it has stood through hardships none of them could bear: the cold rain of winter has sufficed to nourish it, and its faint sun to warm it; the bleak winds have not blanched it, or broken its stem, and the keen frost has not blighted it. Look, Gilbert. It is still fresh and blooming as a flower can be with the cold snow even now on its petals. Will you have it?"

"The Christmas Rose" shows an anonymous poet's wonder at the blooming of the "rose" in winter:

> Know ye the flower that just now blows,
> In the middle of the winter the Christmas rose?
> A plant, indeed, of the crowsfoot kind,
> Not really a rose, but never mind,
> It blooms out o'doors in the garden bed.
> Its petals are white with a tinct of red.
> Though it lacketh perfume to regale the nose,
> To the eyes right fair is the Christmas rose.
> A fiddlestick's end for the frost and snows;
> Sing hey, sing ho, for the Christmas rose.

John Clare, too, marveled at the seasons in "The Winter's Spring":

> I never want the Christmas rose
> To come before its time;
> The seasons, each as God bestows,
> Are simple and sublime.
> I love to see the snow storm hing:
> 'Tis but the winter garb of Spring.

Lines penned in *The Loves of the Plants* (1789) by the great thinker Erasmus Darwin, the grandfather of Charles Darwin, evoke the beauty of the hellebore in winter gardens and its significance in literature and lore:

> Bright as the silvery plume, or pearly shell,
> The snowwhite rose or lily's virgin bell,
> The fair helleborus attractive shone,
> Warmed every sage and every shepherd won.

HEPATICA

The spring is come

Although well known for centuries, the genus *Hepatica* has only ten species. These small perennials are members of the vast Ranunculaceae. They are sometimes called liverleaf and liverworts, but they are not to be confused with the unrelated moss-like, non-flowering, primitive liverworts. Two hepaticas are in North America, *H. acutiloba* with pale pink flowers and *H. americana* with pale blue (rarely white or pink) flowers. The European *H. nobilis* (or *H. triloba*) was thought to be variously an anemone, a wood sorrel, a melilot, and a trefoil.

The genus name is from the Greek word *hepar*, for liver, because of the shape of the plant's lobed leaves, and often their color, suggest a human liver; the same Greek word is the root of the word hepatitis. The Doctrine of Signatures and Nicholas Culpeper advocated the use of hepaticas to "cool and cleanse" the liver as well as to cure yellow jaundice. A diuretic was made from hepatica tea, and an old laxative called Sal Hepatica was once used, but it borrowed only the plant's name— not the plant itself —because this "liver salt" was a patent medicine. Native Americans used hepatica as a tea for a variety of medical complaints and for the curing of bad dreams.

John Parkinson in 1629 described the noble liverwort and attached handsome woodcuts of the plant. He described only the European species, including a double form, and said, "In English you may call them either Hepatica, after the Latine name, as most doe, or Noble Liverwort, which you please." His close observations of hepaticas caused him to remark, "Their diversity among themselves consisteth chiefly in the colour of the flowers."

The English poet John Clare described the first blossoms of the season in "Early Spring" (1860) and used a colloquial name, patty kay, for the hepatica:

> The spring is come, and spring flowers coming, too,
> The crocus, patty kay, the rich heartsease;
> The polyanthus peeps with blebs of dew,
> And daisy flowers; the buds swell on the trees.

In the language of flowers, the hepatica suggested confidence. Nevertheless, some Victorian writers and gardeners seem to have lacked confidence in growing hepaticas. Mrs. C. W. Earle described difficulty with them in *Pot-Pourri from a Surrey Garden* (1898). Later in *More Pot-Pourri from a Surrey Garden* (1899), she related advice from a correspondent on growing them successfully:

> When a child I lived in Somersetshire, where the soil was
> heavy clay. The most beautiful show of Hepaticas I ever saw
> anywhere was a row in an old lady's garden close under a
> thick hedge of Laurestinus with a due north aspect. They
> were single-blue and double-pink. In the same village there
> was for many years a large clump of double-pink close
> under a cottage wall with a south-east aspect. That also
> flowered abundantly, so for double-pink at any rate shade is

not essential, though I remember that the late James Back-
house told me many years ago that the Hepaticas did best
and flowered earliest with a north aspect, as then they went
to sleep sooner in the autumn. The wild ones in Swiss and
French woods are always where they would be shaded in
summer, and grow with the Primroses.

HIBISCUS

The most stately of all herbaceous plants

The genus *Hibiscus* consists of approximately 220 species of annual or perennial shrubs and subshrubs in the Malvaceae. They are native primarily to warm tropical and subtropical regions, being generally tender in other areas. Many cultivated forms of hibiscus are grown as ornamentals or as sources of fiber, like cotton (*Gossypium*), a botanical cousin. Many species of *Hibiscus* are commonly called mallow, but the true mallows belong to the genus *Malva*, and the well-known marsh mallow is *Althaea officinalis*; both are close relatives of *Hibiscus* within the Malvaceae (see the chapter "Marsh Mallow").

Hibiscus militaris, the halberd-leaved rose mallow, grows to 2 meters (7 feet). *Hibiscus moscheutos*, the more hardy common or rose mallow, includes cultivars 'Cotton Candy', 'Crimson', 'Lord Baltimore', and 'Poinsettia'. The rose-of-China, *H. rosa-sinensis*, has a choice form called 'Lateritia Variegata'. Probably the most widely grown in gardens is the tree-like rose of Sharon, *H. syriacus*. *Hibiscus cannabinus* is the Indian hemp or Deccan hemp, probably originating in the East Indies or in Africa. *Hibiscus mutabilis* is known as the Confederate rose in the southern United States, although it is a native of China.

The name *Hibiscus* derives from the old Greek name for mallow, *hibiskos*. The species name *moscheutos* means musk-scented; *mutabilis* means changeable; *rosa-sinensis* means rose of China; and *syriacus* means of Syria, the original home of the plant. In the language of flowers, hibiscus signified delicate beauty.

American naturalist and Quaker William Bartram traveled throughout the southeastern United States recording notes on natural history, the native peoples, and food plants. In his *Travels Through North and South Carolina, Georgia, East and West Florida* (1792), he observed, "The orange flowered shrub Hibiscus is also conspicuously beautiful. It grows five or six feet high . . . . The flowers are of a moderate size, and of a deep splendid yellow." Bartram also saw *Hibiscus coccineus*, the most "stately of all herbaceous plants grow[ing] ten or twelve feet high . . . embellished with large expanded crimson flowers. I have seen this plant of the size and figure of a beautiful little tree, having at once several hundred of these splendid flowers, which may be then seen at a great distance."

The garish red hibiscus native to Hawaii (*Hibiscus kokio*) and similar bright species of other Pacific islands were worn by men: a red hibiscus flower worn behind the right ear suggested, "I am married"; worn behind the left ear, "I am single and looking for a lover"; and worn behind both ears the message was, "I am married but looking for another lover." Regrettably this species is not hardy in the mainland United States or the United Kingdom!

The hibiscus was a loved flower in China, where it was known as the September flower. Poet Ch'en Yun mentioned the hibiscus in "Twilight," translated by Henry H. Hart:

> The night creeps in
> And every sound of human life
> Is hushed.
> Even the tinkle of the camel bell
> Comes muted through the dusk.

In the faint, dying light
Of the waning moon
The first hibiscus flower
Falls.

Chinese poet Mei Sheng of the first century BCE also wrote about hibiscus flowers:

Crossing the river I pluck hibiscus-flowers:
In the orchid-swamps are many fragrant herbs.
I gather them, but who shall I give them to?
My love is living in lands far away.

HOLLY

Come, give the holly a song

Found worldwide as shrubs, trees, and climbers, 400 species of holly comprise the genus *Ilex* in the Aquifoliaceae. The American holly, *I. opaca*; the English or common holly, *I. aquifolium*, consisting of numerous cultivated forms; the Japanese holly, *I. crenata*; and the Chinese holly, *I. cornuta*, are garden evergreens. Deciduous hollies such as winterberry, *I. verticillata*; *I. serrata*; *I. decidua*; and numerous hybrids are now appearing in landscapes. Another holly of gardens, especially in its dwarf form 'Nana', is *I. vomitoria*, or yaupon, an evergreen shrub of the southeastern United States and Mexico.

The generic name for holly is from the Latin name for the European holm oak, *Quercus ilex*, because of the resemblance of the holly leaves to those of the holm oak. The name is preserved in the word *ilicifolius*, for plants with holly-like leaves. The family name Aquifoliaceae derives form *aquifolium*, the classical Latin name for holly, meaning pointy leaves or sharp leaved. Some have suggested that the name is properly spelled *acuifolium*, which has similar meaning. The common name holly comes from *hollen*, the Old English name for the holly tree. In ballads of the British Isles, holly appears as hollin and hollen; from Scotland comes

The ladye walked in yon wild wood
Aneath the hollin-tree,
An' she was aware o' twa bonnie bairns
Were running at her knee.

Edmund Spenser knew the holly as holm or holme. The description from his *Virgils Gnat* (1591) mentions a black holme:

To make the mountains touch the starres divine,
Decks all the forrest with embellishment,
And the blacke holme that loves the watrie vale,
And the sweete cypresse, signe of deadly bale.

In *The Faerie Queene*, Spenser described a grove of trees where a knight and Lady Una stop for protection from a passing storm. The lines catalog the trees that surround them, among which is the holly. Spenser called it the carver holm, since it was favored for carving and turning because of its even grain.

Holly was used by the Romans at the great feast of the Saturnalia, which occurred in mid- to late December to celebrate the winter solstice. Pagan Romans sent evergreen holly sprigs to friends to wish them good health. Later, the season of the Saturnalia was appropriated by Rome's early Christians, who decorated churches and buildings with holly to celebrate Christmas. The practice of decorating with holly dates back much further, however, at least to 1000 BCE and Zoroastrianism, the religion of Persia before its conversion to Islam. Its disciples believed the sun never cast a shadow on the holly, and they sprinkled holly bark water on the faces of newborns to provide good health.

The holly has its share of references in folklore. According to one legend, the crown of thorns made for Jesus by Herod's soldiers was woven of prickly holly leaves. At that time, the plant's berries were yellow; but since the crucifixion, they are red from Jesus's blood.

(Holly berries come in a wide range of colors, though, including white, yellow, and black.) Native Americans used the yaupon holly, *Ilex vomitoria*, to make a restorative tea. Its specific epithet *vomitoria* means to cause vomiting. In South America the mild stimulant drink *yerba maté* is made from *I. paraguariensis*. And in the language of flowers, hollies were for foresight.

An English carol dating from the reign of Henry VI, who reigned in the mid-1400s, praises the holly:

> Nay, ivy, nay it shall not be, I wys,
> Let holly hafe the mastry, as the maner is.
> Holly stand in the halle, fayre to behold;
> Ivy stand without the dore; she is full sore a-cold.

Eliza Cook also sang the praises of holly in "The Christmas Holly":

> The holly! the holly! oh, twine it with the bay—
> Come, give the holly a song;
> For it helps to drive stern winter away,
> With his garments so sombre and long.

The thorns on the holly that arm its leaves were "ordered by an intelligence so wise/As might confound the atheist's sophistries," according to poet laureate Robert Southey of Somerset, who continued in "The Holly Tree" (1813):

> So, serious should my youth appear among
> The thoughtless throng;
> So would I seem, amid the young and gay,
> More grave than they;
> That in my age as cheerful I might be
> As the green winter of the holly-tree.

The holly has long been used as a decoration during the Christmas holiday season. Alfred, Lord Tennyson, wrote of holly: "With trembling fingers did we weave/The holly round the Christmas hearth"

(*In Memoriam*, 1850). In an old Christmas chant called "Modryb Marya
—Aunt Mary," Aunt Mary is the Virgin Mary, mother of Jesus, and
the holly tree with its offering of red berries is a reminder of her:

> Now of all the trees by the king's highway
> Which do you love the best?
> O! the one that is green upon Christmas Day,
> The bush with the bleeding breast.
> Now the Holly with her drops of blood for me:
> For that is our dear Aunt Mary's tree.

The above poem, sometimes listed under the title "Aunt Mary's Tree,"
was composed by Robert Stephen Hawker, the vicar of Morwenstow
in Cornwall. He was known for eccentric dress, usually wearing a
claret-colored coat, a blue fisherman's jersey, and sea boots.

But after the holidays the holly decorations seem out-of-place and
invite superstitions of ill-luck. Robert Herrick ordered the removal
of Christmas decorations on Candlemas Eve, 1 February, to make
room for springtime plants in "Ceremony Upon Candlemas Eve":

> Down with the holly and ivy all,
> Wherewith ye drest the Christmas hall, so
> That the superstitious find
> No one least branch there left behind;
> For look how many leaves there be
> Neglected there—maids trust to me—
> So many goblins ye shall see.

Shakespeare mentioned the holly in act 2 of *As You Like It* (1599), a
comedy with more songs in it than any other of Shakespeare's plays.
In the forest of Arden, to which the rightful duke and his daughter
Rosalind have been banished, Amiens, one of the duke's attending
lords, sings "Blow, blow, thou winter wind":

> Heigh ho! Sing, heigh-ho! Unto the green holly:

> Most friendship is feigning, most loving mere folly:
> Then, heigh-ho! the holly!
> This life is most jolly.

In *The Mystery of Edwin Drood* (1870), Charles Dickens described Christmas Eve at Cloisterham, the fictionalized town of Rochester in Kent, where the verger and his wife are decorating with holly:

> Seasonable tokens are about. Red berries shine here and there in the lattices of Minor Canon Corner; Mr. and Mrs. Tope are daintily sticking springs of holly into the carvings and sconces of the cathedral stalls, as if they were sticking them into the coat button-holes of the Dean and Chapter.

Wordsworth praised the holly grove in "A whirl-blast from behind the hill" (1798):

> Where leafless oaks towered high above,
> I sat within an undergrove
> Of tallest hollies, tall and green;
> A fairer bower was never seen.
> From year to year the spacious floor
> With withered leaves is covered o'er,
> And all the year the bower is green.

One anonymous poet described the dying of the year at the approach of winter and how the holly branches still "shine brightest 'midst the general death":

> From out the hedgerow's faded side,
> Forsaken now by half its pride,
> Still shoots the holly's unchanged green,
> But not in barren beauty seen,
> For, clustered o'er that goodly bough,
> Are scarlet berries blushing now.

   Growing up on farms in eastern North Carolina, my playmates and I made yearly rambles to the woods to collect holly (*Ilex opaca*) for Christmas decorations; we were always looking for branches with the fullest and brightest berries. It is easy to understand the holly's long association with festivals of the winter season because of the plant's glossy foliage and the magnificent show it puts on when the rest of the woods is bare.

HOLLYHOCK

With butterflies for crowns

The sixty species of *Alcea*, known commonly as hollyhocks, are tall-standing plants that sometimes reach a record 7 meters (23 feet). They are herbaceous biennial and perennial members of the cotton, okra, hibiscus, and mallow family, the Malvaceae; all these plants share similarities in the morphology of their flower and seed capsule. It was believed that the hollyhock originated in China, but current investigations find the Middle East or Syria more likely. Regardless of its

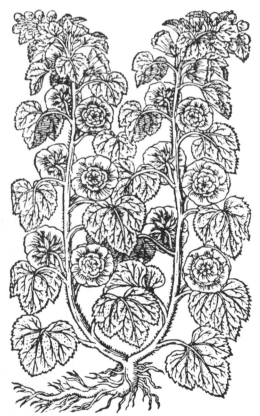

origin, it arrived in English gardens in the 1500s, when it was first reported by William Turner in 1548. The stately hollyhock of gardens is *Alcea rosea* (formerly *Althaea rosea*). Other species are *Alcea ficifolia*, the fig-leaved hollyhock, also known as the Antwerp hollyhock, and the lovely pale yellow *A. rugosa* from the steppes of Russia.

The common name hollyhock is derived from the root words *holy* and *hocc*. The former suggests that it is a blessed or healing herb, not, as some infer, that it reached England via the Holy Land. The latter root word *hocc* is from Old English and means mallow. Two former names for the hollyhock, *Malva Benedicta* and *Caulis Santi*, indicate a legendary origin for the plant related to the lives of saints.

The Latin scientific name for hollyhock, *Alcea*, is derived from the Greek *alkaia*, an old name for the mallow. True mallows are now classified as *Malva* (see the chapter "Marsh Mallow"). Pliny knew the hollyhock in the first century as "a rose growing on a stalk like a mallow." It was called holyoke by William Turner and hollihocke or garden mallow by John Gerard, who noted it also as the "outlandish rose." Similarly, an Old French name from the 1600s is *rose d'outre mer*, the rose from across the sea.

Medicinally the hollyhock has been used as a diuretic, an emollient, and to solve chest problems. In John Gardener's *Feate of Gardeninge* (ca. 1440) hollyhocks are mentioned in a list "of other maner herbys" to plant in the month of April. Some authorities suggest that the name hock was appropriated for this medicinal plant because it was used to reduce the swelling on the hock of a horse's hind leg. Gervase Markham compiled recommended husbandry techniques for raising farm animals in *Cheape and Good Husbandry for the Well-Ordering of all Beastes and Fowles* (1614) and suggested, "Annoint her feet with the juyce of the Hearb Holyhocke."

In the language of flowers hollyhocks suggested fruitfulness or fecundity, perhaps alluding to their abundance of both flowers and seeds. The yellow hollyhock became a recommended church decoration for 29 August, the day to commemorate St. John the Baptist's death.

Early botanists and poets took note of the hollyhock. In 1626 Francis Bacon mentioned hollyhocks in *Sylva Sylvarum*: "This Experiment of Severall colours, comming up from one Seed, would be tried also in . . . Poppy and Hollyoke." In 1700, Abraham Cowley wrote in *Of Plants*: "The Holihock disdains the common size Of Herbs, and like a Tree do's proudly rise." Tennyson noted in 1830 in "Song: A Spirit Haunts," a poem about the year's last hours: "Heavily hangs the hollyhock,/Heavily hangs the tiger lily." Victorian poet Jean Ingelow described "Queen hollyhock, with butterflies for crowns" in part I of "Honors." Robert Louis Stevenson wrote *A Child's Garden of Verses* in which is "The Flowers":

All the names I know from nurse:
Gardener's garters, shepherd's purse;
Bachelor's buttons, lady's smock,
And the lady hollyhock.

Unkist, Unkind! (1897) by Violet Hunt is a gothic tale that explores the sexual frustration and loneliness of women who are "only the companion—poor, plain, and ill connected" of their husbands and bachelor boyfriends. One of the characters, Lady Darcie, is staying at Swinope House, a guest house situated on a "desolate expanse of moorland" in Northumberland. She longs to be in London, away from the bleak landscape, and stares forlornly out the window, saying,

> "This kind of country does not go well with me. It is not at all a good background, and I always like to choose my background colors carefully! These colours," pointing to the delicate folds of her pale green and pink morning gown, "don't go well with a country where all the high lights are steel colour, and all the shadows black. It takes a great bouncing hollyhock of a woman to look well here, not a white lily, as they call me in town."

A story from England's west country credits fairies with the origin of hollyhocks. According to legend, each year on midsummer's day (around 21 June) an island appeared where the Wye and Severn rivers join at the border of Wales and Gloucestershire. The island was full of flowers, trees, and singing birds, and in the middle was a castle inhabited by fairies. Mortals on shore were allowed to visit the fairy island by going through a tunnel under the river. During the visits the fairies were never seen, but lively music could be heard as invisible hands provided food and drink. The one rule in order to continue to enjoy the hospitality of the invisible fairies was never to take anything from the island. One year, however, a girl wanted to return home with a bouquet of flowers she had picked on the island. Her mother forbade her, but the naughty girl kept one blossom and hid it in her pocket. As the mother led her child through the tunnel, the girl was turned into a hollyhock, her pink pinafore becoming pink blossoms. Never again did the fairy island appear on midsummer's day.

American writer Celia Thaxter gardened on Appledore, one of the Isles of Shoals off the coast of Portsmouth, New Hampshire, and wrote about her experiences in *An Island Garden* (1894). In a collection of poems primarily about flowers and her tiny garden, she fretted about the coming frost and the loss of her hollyhocks in "My Hollyhock":

> Ah me, my scarlet hollyhock,
> Whose stately head the breezes rock,
> How sad, that in one night of frost
> Thy radiant beauty shall be lost,
> And all thy glory over thrown
> Ere half thy ruby buds have blown!

The bold hollyhock is short-lived, and it is generally grown as an annual or at most a biennial. It is the plant my grandmother grew beside the smokehouse. Its popularity has waned and it is now considerably less grown than in the nineteenth and early twentieth centuries. But still the grand and stately flowers provide midsummer elegance to gardens with the space to accommodate them.

HONEYSUCKLE

Untouched thy honeyed blossoms blow

One hundred eighty species of honeysuckle are native primarily to the Northern Hemisphere and belong to the genus *Lonicera* in the Caprifoliaceae. Honeysuckle plants, also called woodbine, may be evergreen or deciduous, climbing, twining, creeping, or even bushy. At least a few of the species are rampant and will become weedy; however, numerous species and their cultivars are useful garden ornamentals. Well known in Europe are *Lonicera caprifolium*, the Italian woodbine, and *L.*

periclymenum, the common honeysuckle. North Americans know the naturalized *L. japonica*, Japanese honeysuckle, and the native evergreen *L. sempervirens*, coral or trumpet honeysuckle.

Honeysuckle was given the botanical name *Lonicera* by Linnaeus in honor of Adam Lonitzer who lived in the sixteenth century. Lonitzer was a German botanist and naturalist who wrote the popular herbal *Kreuterbuch* in 1557. The species name *periclymenum* comes from an old Greek name for the honeysuckle. The name *japonica* suggests "from Japan," and *sempervirens* means evergreen. The name *caprifolium*, like that of the family, means goat leaf, suggesting that the leaves were used as fodder for goats. Honeysuckle retains the name of goat or goat's leaf in French, *chèvre-feuille*, German, *Geis-*

blatt, and Italian, *capri-foglio.* The English name honeysuckle comes from *honi* and *sucan,* Old English words referring to the pleasure of sucking honey from the corolla tube. In the thirteenth century, *hunisuccle* also referred to clover and its flowers (*Trifolium*). In the language of flowers the common honeysuckle meant sweetness of disposition because of the sweet scent of the flowers. It also implied a bond or meant "captive of love," suggested by the plant's twining growth habit that embraces trees and other plants. The common name woodbine comes from Middle English and refers to the ability of the plant to tie or bind as it grows and climbs.

John Dryden, a prolific writer of the seventeenth century, in 1697 produced the finest translation of Virgil's *Georgics.* It mentions honeysuckle: "Then Melfoil beat, and Honey-suckles pound, With these alluring Savours strew the Ground."

In early English writing, the woodbine appears in part 2 of Chaucer's *Knight's Tale* (ca. 1387): "To maketh him a garland with a spray / Of woodbine leaves." And in *Troilus and Criseyde,* Chaucer's longest complete poem, the clinging and twining of the woodbine is an example followed by the lovers:

> Whan she his trouthe and clene entente wiste,
> And as aboute a tree, with many a twiste,
> Bitrentand wryth the sote wode-binde,
> Gan eche of hem in armes other winde.

Shakespeare wrote in *A Midsummer Night's Dream* (1595) of a place where Titania, queen of the fairies, sleeps. Oberon, king of the fairies, knows this place, with its bank

> where the wild thyme blows,
> Where oxlips and the nodding violet grows,
> Quite over-canopied with luscious woodbine,
> With sweet musk-roses and with eglantine.

Later in act 4, Titania, under the spell of a love potion, tells Bottom, a weaver and the leader of a group of blundering traveling tradesmen:

> Sleep thou, and I will wind thee in my arms.
> Fairies, be gone, and be all ways away.
> So doth the woodbine the sweet honeysuckle
> Gentle entwist; the female ivy so
> Enrings the barky fingers of the elm.
> O, how I love thee! how I dote on thee!

Shakespeare mentioned honeysuckle again in act 3 of *Much Ado About Nothing* (1598):

> the pleached bower,
> Where honeysuckles, ripen'd by the sun,
> Forbid the sun to enter, like favorites
> Made proud by the princes.

A few lines later, Beatrice is hiding in a tangle of vines, and Ursula says to Hero in forming a scheme, "So angle we for Beatrice; who even now is couched in the woodbine coverture. Fear you not my part of the dialogue." And in book 9 of John Milton's *Paradise Lost* (1667), Eve tells Adam of the plants, herbs, and flowers "still to tend" and of the pruning and propping to be done:

> Let us divide our labors; thou where choice
> Leads thee, or where most needs, whether to wind
> The woodbine round the arbor, or direct
> The clasping ivy where to climb; while I
> In yonder spring of roses intermixed
> With myrtle find what to redress till noon.

English writers of the eighteenth and nineteenth centuries continued to make reference to honeysuckle. Robert Tannahill was a hand-loom weaver who wrote on the side. He was so despondent over the harsh criticism of his poems and writings that he drowned himself in a culvert in Paisley. His poem "The Midges Dance Aboon the Burn" describes birds, trees, and flowers that were dear to him:

> The honeysuckle and the birk
> Spread fragrance through the dell.
> Let others crowd the giddy court
> Of mirth and revelry
> The simple joys that nature yields
> Are dearer far to me.

Robert Burns, who wrote equally well in eighteenth-century English and in his native Scottish, knew the honeysuckle as woodbine. In "The Banks o' Doon," he wrote,

> Aft hae I roved by bonie Doon,
> To see the rose and woodbine twine;
> And ilka bird sang o' its luve,
> And fondly sae did I o' mine.

John Keats wrote of "Meg Merrilies," the gypsy who lived on the moors:

> But every morn of woodbine fresh
> She made her garlanding,
> And every night the dark glen Yew
> She wove, and she would sing.

William Wordsworth also wrote of honeysuckle in "White Doe of Rylstone" (1815), in which Emily, "the heart-sick maid,"

> approached yon rustic shed,
> Hung with late-flowering woodbine, spread
> Along the walls and overhead
> The fragrance of the breathing flowers
> Revived a memory of those hours.

Denis Florence MacCarthy described honeysuckle in his depiction of "Summer Longings," also published as "Waiting for the May":

> Waiting for the pleasant rambles
> Where the fragrant hawthornbrambles
> With the woodbine alternating,
> Scent the dewy way.

Coventry Patmore, who worked in the book department of the British Museum after his father's financial collapse, wrote in "Arbor Vitae" that the tree was "With honeysuckle, oversweet, festoon'd." West Country cleric Richard Graves wrote in "Euphrosyne, or Amusements on the Road of Life" (1776): "See! how that woodbine round the door/And lattice blooms!"

New York–born Philip Freneau, often called the poet of the American Revolution, wrote volumes of political essays, satires, and nature poems, which were considered a precursor of the Romantic Movement. In 1786 he wrote "The Wild Honeysuckle":

> Fair flower, that dost so comely grow,
> Hid in this silent, dull retreat,
> Untouched thy honeyed blossoms blow,
> Unseen thy little branches greet:
> No roving foot shall crush thee here,
> No busy hand provoke a tear.

Poet and painter Dante Gabriel Rossetti wrote "The Honeysuckle," showing the extra effort he found worthwhile just to hold the sweet flower:

> I plucked a honeysuckle where
> The hedge on high is quick with thorn,
> And climbing for the prize, was torn,
> And fouled my feet in quag-water;
> And by the thorns and by the wind

The blossom that I took was thinn'd,
And yet I found it sweet and fair.

Captain John Smith, explorer of North America, writing in *The Generall Historie of Virginia, New England and the Summer Isles* (1624), described "A kinde of Wood-bind . . . which runnes upon trees, twining it selfe like a Vine: the fruit . . . eaten worketh . . . in the nature of a purge."

English novelists used honeysuckles in domestic descriptions. In Jane Austen's *Sense and Sensibility* (1811), her first published work, readers see the heroine Elinor's home in Devonshire, which had been offered to the Dashwoods by Sir John Middleton: "As a house, Barton Cottage, though small, was comfortable and compact; but as a cottage it was defective, for the building was regular, the roof was tiled, the window shutters were not painted green, nor were the walls covered with honeysuckles." And in *Pickwick Papers* (1836), Charles Dickens wrote, "There was a bower at the further end, with honeysuckle, jessamine, and creeping plants—one of those sweet retreats which humane men erect for the accommodation of spiders."

Some species of the honeysuckle, or woodbine, emit a sweet fragrance from their trumpet-shaped flowers while they seductively embrace and sometimes smother other plants. English diarist Samuel Pepys noted that these "ivory bugles blow scent instead of sound."

HYACINTH

If spring proves mild, 'tis hyacinthus time

Hyacinths are bulbous plants that have undergone various taxonomic changes in the last few decades. Currently, plants commonly called hyacinth are scattered over several genera, including *Hyacinthus, Hyacinthoides, Scilla, Bellevalia,* and *Brimeura.* The name *Endymion,* which in Greek mythology was the name of the beautiful youth beloved by Diana, has fallen out of favor. Three species are currently recognized in the genus *Hyacinthus.* They are in the Liliaceae (or the smaller, more recently defined Hyacinthaceae) and were historically distributed in western and central Asia. The one species that is mostly widely known and grown is *Hyacinthus orientalis,* the common hyacinth. The two other species are *H. litwinowii,* from Iran and Russia, and *H. transcaspicus,* also from Iran. The genus name *Hyacinthus* is from the Greek name *hyakinthos,* used by Homer, Ovid, and others for a flowering plant that is unlikely to have been the modern hyacinth. The hyacinth of classical literature has been taken to be various plants including gladiolus, martagon lily, fritillary, iris, and larkspur. In Elizabethan English, the name jacinth or jacynth would sometimes appear for hyacinth. The Greeks also applied the name jacinth to a precious blue stone, probably the sapphire or aquamarine; thus the classic hyacinthine color is blue. In the Victorian language of flowers hyacinths suggested sport, game, and play; the purple hyacinth meant sorrow; and the blue hyacinth was the symbol for constancy and fidelity.

From Greek mythology, the hyacinth is a symbol

of rebirth and the renewed plant growth of spring; it is also a symbol for youthful male beauty. Hyacinthus, son of the Spartan King Amyclas, was an outstandingly handsome young prince beloved by the sun god Apollo, called Phoebus in Roman mythology, the son of Zeus. Zephyr, the god of the west wind, was jealous of Apollo's affections for the prince. One day when Hyacinthus joined Apollo in discus throwing, Zephyr intentionally blew Apollo's stone plate off course and sent it crashing into the skull of poor Hyacinthus, killing him on the spot. From his blood sprang the plant that is named in his memory. As Ovid told the legend, "One day, near noon . . . they both stripped off their clothes and oiled their limbs, so sleek and splendid, and began the game, throwing the discus." And when the discus struck Hyacinthus, "the wound was past all cure." Milton also wrote of the flower and young Hyacinthus in *On a Death of a Fair Infant Dying of a Cough* (1626). He likened the death of "so sweet a child" to that of Hyacinth:

> Yet art thou not inglorious in thy fate;
> For so Apollo, with unweeting hand,
> Whom did slay his dearly-loved mate,
> Young Hyacinth, born on Eurotas' strand,
> Young Hyacinth, the pride of Spartan land,
> But then transformed him to a purple flower:
> Alack that so to change thee Winter hath no power.

The myth of Apollo and Hyacinthus retained its significance for centuries. The boy was worshipped in early Crete as the god of flowers, and in Sparta an annual spring celebration called Hyacinthia or the Hyacinthus Festival commemorated his death. The festival was a three-day affair of grieving for the fallen Spartan. Over time it evolved into a celebration of the might and power of Apollo. A related legend originating from the abrupt arrest of Hyacinth's young life claims that eating bulbs of the plant will delay the onset of adulthood, retarding the growth of facial hair and other signs of puberty.

It is unlikely, however, that the classical flower in the legend was a hyacinth; some think it was *Gladiolus italicus*. The much disputed true plant supposedly reveals the Greek letters AIAI, for "Alas, alas," a mournful exclamation uttered by Apollo at the death of Hyacinthus. The ancients thought they could decipher these letters of grief on the petals.

In 1578 Henry Lyte made an early use of hyacinth in English in his translation of Rembert Dodoens's *Niewe Herball or Historie of Plantes*: "Of the redde Lillie Ouide [Ovid] wryteth this, that it came of the bloud of the Boy Hyacinthus . . . . And for a perpetuall memorie of the Boy Hyacinthus, Apollo named these floures Hyacinthes." In "Lycidas," Milton called the hyacinth that "sanguine flower" and referred to the legend of the inscribed petals:

> Next Camus, revered sire, went footing slow,
> His mantle hairy, and his bonnet sedge,
> Inwrought with figures dim, and on the edge
> Like that sanguine flower inscribed with woe.
> Ah; who hath reft (quoth he) my dearest pledge
> Last came and last did go.

Camus is the god of the River Cam, a deity of significance to Cambridge University. René Rapin also told the story of Hyacinthus and Apollo:

> If spring proves mild, 'tis Hyacinthus time,
> A Flower which also rose from Phoebus' crime;
> The unhappy Quoit which rash Apollo threw,
> Obliquely flying, smote his tender Brow.
> And pale alike he fell, and Phoebus stood,
> One pale with Guilt, and one with loss of Blood;
> Whence a new Flower with sudden Birth appears,
> And still the Mark of Phoebus' Sorrow wears;
> Spring it adorns, and Summer's scenes supplies
> With blooms of various Forms and various Dyes.

Percy Bysshe Shelley, writing in *Adonais* (1821), his elegy to John Keats, described his grief by imagining that of Apollo at the loss of Hyacinth and that of Aphrodite at the loss of Adonis. Like Hyacinth, Adonis was a beautiful youth killed prematurely through an act of jealousy. Shelley wrote,

> Grief made the young Spring wild, and she threw down
> Her kindling buds, as if she Autumn were,
> Or they dead leaves; since her delight is flown
> For whom should she have waked the sullen year?
> To Phoebus was not Hyacinth so dear,
> Nor to himself Narcissus, as to both
> Thou, Adonais; wan they stand and sere
> Amid the faint companions of their youth,
> With dew all turned to tears; odor, to sighing ruth.

Milton also referred to Adonis. In part 4 of *Comus* (1634), a pastoral drama, the attendant spirit describes "odorous banks that blow / Flowers of more mingled hue" and

> Beds of Hyacinth and roses,
> Where young Adonis oft reposes,
> Waxing well of his deep wound
> In slumber soft, and on the ground.

The hyacinth was often used as a herald of spring. Ebenezer Elliott of Sheffield wrote poems of poverty and oppression and was known as "the Corn Law Rhymer" because he denounced the old corn laws dating from 1360 which regulated export and import of grain, resulting in hardships to the poor when bread prices rose after poor harvests. Elliott described the spread of spring plants "o'er every hill" in his poem "Spring" (1846): "Thy leaves are coming, snowy-blossomed thorn, / Wake, buried lily! Spirit, quit thy tomb! / And thou shade-loving hyacinth, be born!" Shelley wrote in "The Sensitive Plant" (1820) of

> The hyacinth purple, and white and blue,
> Which flung from its bells a sweet peal anew
> Of music so delicate, soft, and intense,
> It was felt like an odour within the sense.

And Tennyson's epic twelve-part poem *The Idylls of the King* (1833), which presents the story of King Arthur, also uses the hyacinth to mark the spring. Queen Guinevere is hiding in a nunnery at Almesbury after her adulterous affair with Sir Lancelot. A novitiate befriends her, prompting Guinevere to recall the "golden days" in which she first saw the "best knight and goodliest man":

> for the time
> Was may-time, and as yet no sin was dream'd,—
> Rode under groves that look'd a paradise
> Of blossom, over sheets of hyacinth
> That seem'd the heavens upbreaking thro' the earth.

Hyacinths are sometimes called bluebells, but this common name leads to a bit of confusion because of other well-known bluebells in different genera. For example, the famous bluebells of Scotland are in the unrelated genus *Campanula* (see the chapter "Bellflower"). More closely related are the bluebells of *Hyacinthoides*, a genus including species once classified as *Scilla* and *Endymion*. In the Liliaceae (or the smaller, more recently defined Hyacinthaceae), the three bulbous species of *Hyacinthoides* grow in a range of habitats from woodlands, to seasides, to mountainous areas up to elevations of 1500 meters (4900 feet). All are native to northern Africa and western Europe, and their flowers are blue to blue-violet.

The larger-leafed Spanish bluebells are *Hyacinthoides hispanica*; *H. italica* has deep blue, starry flowers; and the woodland English bluebells, also called wild or wood hyacinth, are *H. non-scripta*. *Hyacinthoides non-scripta* illustrates the close relationship between the genera *Hyacinthus* and *Hyacinthoides*; the specific epithet, chosen by Linnaeus as *non-scripta*, means not inscribed or without markings, because he could not find the Greek AIAI written on the flower as it should have been according to the legend of Apollo's despairing cry at the death of Hyacinthus.

The bluebells and wood hyacinths of *Hyacinthoides* are well known as a wild flower of the British Isles. In the nineteenth century they appeared in numerous places in English literature. Robert Southey—a leader in the poetry of the time through his association with the Lake Poets, a group that included Coleridge and Wordsworth—wrote in 1801 in *Thalaba, the Destroyer*, "Amid the growing grass/The blue-bell bends, the golden king-cup shines,/And the sweet cowslip scents the genial air." In 1851 in *Sketches of Natural History*, poet Mary Botham Howitt described "The nodding Bluebell's graceful flowers,/The Hyacinth of this land of ours."

Eliza Cook, a self-educated Victorian poet from London whose writings are characterized by domestic sentimentality, chose a bed of bluebells above all others in "Blue Bells in the Shade":

> The choicest buds in Flora's train let other fingers twine;
> Let others snatch the damask rose, or wreathe the eglantine;
> I'd leave the sunshine and parterre, and seek the woodland
>   glade,
> To stretch me on the fragrant bed of blue bells in the shade.

And Francis Kilvert, a vicar in Wiltshire and Wales, wrote in his diary of such a serene place filled with azure bells:

> WEDNESDAY 17 MAY 1876. Over the gate of the meadow there leaned a beautiful wild cherry tree, snowy with blossom, that scented the air far and wide. And along the wild broken bank and among the stems of the hawthorn hedge there grew a profusion of bluebells. I never saw bluebells more beautiful. They grew tall and stately, singly or in groups, and sometimes in such a crowd that they filled the

hollow places and deep shadows of the overarching hedge
with a sweet blue gloom and tender azure mist among the
young bright fern. Here or there a sunbeam found its way
through a little window or skylight in the thick leafage overhead and singling out one bluebell amongst the crowd
tipped the rich and heavily hanging cluster of bells with a
brilliant azure gleam and blue glory, crowning a flower a
queen among her ladies and handmaidens who stood
around in the background and green shade.

At least one writer addressed the issue of the somewhat mislead-
ing application of common names to *Hyacinthoides* and its relatives.
The *London Daily News* of 30 June 1897 noted that "Sir Herbert Max-
well objects to the southron use of the name bluebells, as applied to
the flowers that he prefers to call wood hyacinths."

IRIS

O flower-de-luce, bloom on

Amazingly, the rainbow-colored blossoms of the 300 species of *Iris* are exclusively distributed in the Northern Hemisphere. To organize the multitude of species and cultivars, the genus has been classified into twenty-seven subgenera. In addition, botanical cousins of the iris form nearly fifty different genera in the Iridaceae. The iris is one of the oldest cultivated plants. Favorite garden forms include the much-hybridized bearded iris; two diminutive species, *I. cristata* and *I. reticulata*; the Japanese water iris, *I. ensata*; the native iris of the Pacific coast of the United States and the Louisiana hybrids; the roof iris, *I. tectorum*; *I. variegata*; and the beloved winter iris, *I. unguicularis*. All are either rhizomatous or bulbous perennials.

An unknown herbalist wrote about *Iris variegata*: "I am not able to express the sundry colours and the mixtures contained in the flower; it is mixed with purple, yellow, blacke, white, and fringe of thrum down the middle of the lower leaves, of a whitish yellowe . . . . Tipped or fringed, and as it were raised up of a deepe purple colour near the ground." Of another form of iris he said, "the whole flower of the colour of a ginnie henne, and a rare and beautiful flower to behold."

Because of the diverse colors found among its flowers, the plant is named for the Greek goddess Iris, the divine personification of the rainbow and a swift-footed messenger. Iris glided along the rainbow to the ends of the earth and to the underworld carrying messages to the gods and to men. The language of flowers took the iris's meaning from its namesake: the flower signaled that the giver had a message for the recipient. In Homer's *Iliad* Iris was the one who "runs on rainy wind" and carries a message from Zeus to the gods that they should not encourage the Trojans. Another of Iris's duties was to lead the souls of women to the Elysian fields (Mercury performed this task for the men); as a result Greeks often put iris blossoms on the graves of women. Plutarch wrote that iris is "the eye of heaven," a meaning that encompasses both the rainbow and the colored part of the human eye. Also, many believe it is the likeness of the iris that adorns the brows of the Egyptian Sphinx.

Various stories and legends, some based partially on historical facts, associate the iris with the French monarchy. The iris figured in Louis VII's unsuccessful crusade in 1147, in which he adopted the purple iris as his emblem because of a dream. From this association arose the name fleur-de-Louis, which developed into fleur-de-luce and then fleur-de-lis. Appearing on the French flag, the three parts of the fleur-de-lis, or "flower of the lily," represented faith, wisdom, and valor. In another legend explaining the adoption of the iris as a national symbol, Clovis I, the first Christian king of France, in battle in CE 496, found himself trapped between a lake or river and the opposing Germanic army. Praying to escape safely, he saw in the distance yellow flag irises growing in the water. He realized that the water must not be too deep if the irises could grow there, and he led his army safely across. Clovis declared the yellow iris, probably *I. pseudacorus*, the symbol of his army's salvation and replaced the toads on his banner with the tripartite flower. In another version of Clovis's legend, the purple iris was his salvation. The iris is also known as the fleur-de-luce, or the "flower of light," said to have been brought from heaven to Clovis. In the *Boke of St. Alban*, the anonymous author noted that "the armes of the King of France were certainle sende by an awngell from heaven, that is to say iii [three] flowers in manner of swerdis [swords] in a felde of azure, the wich certain armys were gewyn to the forsayd Kinge in sygn of everlasting trowball, and that he and his successors alway with bataill and swerddys sholde be punyshid." Finally, one French legend claims the purple iris is a gold-colored flower that put on the purple color and went into everlasting mourning on the night of the crucifixion of Jesus.

The roots of some iris, such as, *Iris pallida*, the Dalmatian iris, have been used to make a powder called orrisroot, which name itself is a variant of "iris." It is used in cosmetics, in orange-flavored liqueurs, in sachets, and in tooth powders. It provides the odor of violets and has fixative qualities that reduce evaporation. Orris was known to the Greek physician Dioscorides in the first century. He claimed that it could cure numerous ailments: stomachache, venomous bites, sores, and the "humors." In North America, Native Americans recognized the medicinal properties of the native irises. William Bartram wrote in his *Travels Through North and South Carolina, Georgia, East and West Florida* (1792) of seeing a tribe of Ottasses Indians from the area of Georgia growing a "small plantation" of *Iris versicolor* for use as a physic or purgative.

Literary references to the iris began to appear as early as Chaucer's writings in the fourteenth century. In the *Prologue* to *The Canterbury Tales* (1387–1400) the traveling company is introduced, including the friar, who wins a prize for his renowned ballad-singing abilities:

> Wel coude he singe and pleyen on a rote.
> Of yeddings he bar utterly the prys,
> His nekke whyt was as the flour-de lys;
> Ther-to he strong was as a champioun.

The yellow iris is often called the flag iris or the corn-flag, an old name shared with gladiolus. Ben Jonson wrote, "Bring corn-flag, tulip and Adonis flower/Fair ox-eye, goldy-locks, and columbine,/Pinks, gowlans, king-cups and sweet sops-in-wine." And Edmund Spenser, who is buried near Chaucer in Westminster Abbey, wrote in *The Faerie Queene*, "The lilly, Ladie of the flowering field/The Flower-deluce, her lovely Paramoure."

Shakespeare knew of the messenger Iris, alluding to her in *Henry VI, Part II* (1592). Queen Margaret tells a departing guest: "Let me hear from thee:/For whereso e'er thou art in this world's globe,/I'll have an Iris that shall find thee out." In *Henry VI, Part I*, Shakespeare used the iris flower in a metaphor. At the funeral of Henry V at Westminster Abbey during the wars with France, a messenger from France enters and cries,

> Awake, awake, English nobility!
> Let not sloth dim your honours new-begot
> Cropp'd are the flower-de-luces in your arms;
> Of England's coat one half is cut away.

René Rapin described the varied blooming of the iris, referring also to the origin of its name:

> Fair Iris now an endless Pomp supplies,
> Which from the radiant Bow that paints the skies,
> Draws her proud Name and boasts as many Dyes,
> For she her Colour varies and her kind
> As ev'ry Season to her Growth's inclin'd.

The English poet of the early nineteenth century John Clare referred to the iris simply as flags in "Farewell":

> Farewell to the bushy clump close to the river
> And the flags where the butter-bump hides in forever;
> Farewell to the weedy nook, hemmed in by waters;
> Farewell to the miller's brook and his three bonny daughters.
> Farewell to them all while in prison I lie—
> In the prison a thrall sees naught but the sky.

In the same era, Henry Wadsworth Longfellow, American poet and professor, wrote about "The Iris":

> O flower-de-luce, bloom on, and let the river

Linger to kiss thy feet!
O flower of song, bloom on, and make for ever
The world more fair and sweet.

Christina Rossetti likewise appreciated the fair irises, as she mentioned them in "Reflection":

I have strewn thy path, beloved,
With plumed meadowsweet,
Iris and pale perfumed lilies,
Rose most complete;
Wherefore pause for listless feet?

In Rossetti's "Birthday" (1857) the fleur-de-lis is something special, something extra for the singular day: "Raise me a dais of silk and down; …/ Work it in gold and silver grapes/ In leaves and silver fleurs-de-lys."

Ralph Waldo Emerson wrote in "Hamatreya" of the naturalness of the iris, feeling in them a connection with his own natural place in the world:

How graceful climb those shadows on my hill!
I fancy these pure waters and the flags
Know me, as does my dog: we sympathize;
And, I affirm, my actions smack of the soil.

Walt Whitman's "Song of Myself" includes flag iris in much the same way. The plant's flourishing in the North American soils parallels his own:

Breast that presses against other breasts it shall be you!
My brain it shall be your occult convolutions!
Root of washed sweet flag! timorous pond snipe!
Nest of guarded duplicate eggs! it shall be you!
Mixed tussled hay of head, brawn, it shall be you!
Trickling sap of maple, fiber of manly wheat, it shall be you!

In my home in North Carolina, spring comes intermittently in the mountains. Days of sun alternate with days of snow as late as April in the Blue Ridge, where *Iris cristata* grows along the road leading to Mt. Jefferson, an elevation of 1500 meters (5000 feet). I think of Iris, the messenger, when I see a rainbow after the passage of a violent spring storm and when the valley awakes from its winter chill to a new yellow-green. Summer is on its way.

IVY

Creeping on where time has been

The ivy plant, *Hedera*, in the Araliaceae, is native to Europe, Asia, and northern Africa and consists of eleven species. Of all garden plants, they are among the most well known—not for their flowers, which are small and relatively inconspicuous, but for their foliage. The common or English ivy, *H. helix*, has scores of cultivated forms, such as 'Oro di Bogliasco', known in North America as 'Goldheart', with a strikingly brilliant center of yellow; 'Ivalace' with margins so undulate as to appear filigreed; and the white-margined and marbled 'Glacier'. Additional species of the garden or landscape include *H. canariensis* (formerly *H. algeriensis*), the Algerian ivy, which has a handsome variegated cultivar 'Gloire de Marengo', and the bold-leaved *H. colchica* from Asia Minor. All the species are woody evergreen plants that climb or creep. Ivies have two growth stages: the sterile juvenile climbing stage that has primarily hand-like lobed leaves and the less common fertile adult tree-like stage that has simple leaves. The size of ivy leaves varies considerably, from the size of a person's hand to miniature-leaved plants capable of being tucked into an alpine or rock garden.

The name *Hedera* is the classical Latin name for ivy that Linnaeus retained for his binomial naming scheme, *Species Plantarum* (1753). The common English name ivy is derived from the Old English word *ifig*, which has related names in other Germanic languages. It is possible that the English word ivy is distantly related to the Latin *ibex*, which literally means climber, like the goat of the same name. The species name *helix* means "to wind or climb around," possibly referring to the staff entwined with ivy carried by Bacchus, or Dionysus in Greek mythology.

Bacchus, the god of wine, revelry, and fertility, was associated with ivy in Roman mythology. He is usually represented as a beautiful youth with dark eyes, golden hair with flowing curls, and ivy twining around his head and shoulders. Bacchus is sometimes shown holding grapes and driving a chariot drawn by panthers. Various versions of the story of Bacchus tell how he came to be associated with ivy. Bacchus was the son of Jupiter, or Zeus, and the mortal Semele. When Semele asked Jupiter to appear to her in all his divinity, he did, and she was destroyed in a blaze of glory. In some versions, her child was yet unborn but was not destroyed along with her; the baby was sewn into Jupiter's thigh to finish its development. In other versions, Bacchus was sheltered in an ivy bush until he grew up. Bacchus worshipped the ivy plant, which he called *kissos*, the Greek name for ivy and the name given to the infant Bacchus by Semele.

In keeping with the influence of Bacchus, in the Middle Ages a bunch of ivy hanging on a pole in front of a building, called an alepole or stakepole, indicated the presence of a tavern. Though first associated with pagan practices, the ivy eventually became a part of the Christian Christmas celebration through its use as a traditional decoration alongside the holly. A medieval carol praises it:

> Ivy is green, with coloure bright,
> Of all trees best she is,
> And that I prove will be right,
> *Veni coronaberis.*

In seeming opposition to the influence of Bacchus, the language of flowers suggested that ivy signaled fidelity, marriage, and friendship. If the ivy had a new sprig of growth, it meant, "I am eager to please." Victorians bought friendship brooches showing an ivy clinging around a tree, with an inscription in English or Latin: "Nothing can detach me from you."

Botanically ivy was described in the earliest texts. Virgil recommended in *Georgics* (30–37 BCE) the best soil types for different plants and gave instructions for making classifications. John Dryden's translation of 1697 instructs:

> 'Tis easy to distinguish by the sight,
> The colour of the soil, and black from white.
> But the cold ground is difficult to know;
> Yet this, the plants that prosper there will show—
> Black ivy, pitch trees and the baleful yew.

Ancient Greek writers incorporated ivy into their poetry. In "The Cup" Theocritus wrote, "About its lip winds ivy—ivy flecked/With golden berries, which goes twisting round/In all the glory of its saffron fruit." A yellow-berried form of of *H. helix* is var. *poetarum*, a name meaning "of the poet." And at about the same time, ca. 300 BCE, Simias wrote "Sophocles's Tomb," translated by Walter Leaf, in which he implored,

> Wind gently, ivy, o'er the tomb,
> Gently, where Sophocles is laid;
>
> . . . .
>
> Twine thy lithe tendrils, gadding vine,
> To praise the cunning of his tongue.

One of the earliest English literary references to ivy is in Chaucer's *Palamon and Arcite*, later adapted as *The Knight's Tale* in *The Canterbury Tales* (1387–1400). Theseus referees an argument between Arcite and Palamon, two suitors quarreling over his sister Emelye. Theseus reminds them that she cannot wed them both:

> Ye know yourselves, she cannot marry two
> At once, though ye eternal combat do:
> One of you, be he glad or full of grief,
> Must go and whistle to an ivy leaf.

In Edmund Spenser's *Shepheardes Calender*, a writing of the sixteenth century, Thomalin goes hunting with a bow one holiday in March (a "Yuie todde" is an ivy bush):

> At length within an Yuie todde
> (There shrouded was the little God)
> I heard a busie bustling,
> I bent my bolt against the bush,
> Listening if any thing did rushe,
> But then heard no more rustling.

In Alexander Pope's "Spring, the First Pastoral" (1709), Daphnis, a shepherd of Greek mythology considered the inventor of pastoral poetry, and who died of unrequited love, contemplates a bowl inscribed with images of the seasons:

> And in this bowl, where wanton ivy twines,
> And swelling clusters bend the curling vines;
> Four figures rising from the work appear,
> The various seasons of the rowling year.

In the eighteenth century English naturalist Gilbert White recorded notes on the ivy in his journals, which revealed his insatiable curiosity for animal and plant life:

> SEPT. 18 [1772] Ivy begins to blow: & is the last flower which supports the hymenopterous, & dipterous, Insects. On

sunny days, quite on to Nov. they swarm on trees covered
with this plant; & when they disappear probably retire
under the shelter of it's leaves, concealing themselves
between it's fibres, & the tree that it entwines.

In part 7 of "The Rime of the Ancient Mariner" (1798), Samuel
Taylor Coleridge described the wood near the sea where an old hermit lives and talks with the mariners:

> Brown skeletons of leaves that lag
> My forest-brook along;
> When the ivy tod is heavy with snow,
> And the owlet whoops to the wolf below,
> That eats the she-wolf's young.

And in the early nineteenth century Felicia Dorothea Hemans
wrote "The Grave of a Poetess" in reference to the village churchyard
of Kilkenny where the poet Mary Tighe is buried:

> Fresh leaves were on the ivy bough
> That fringed the ruins near;
> Young voices were abroad, but thou
> Their sweetness couldst not hear.

The ivy has been given a place in the poetry of love. Elizabeth Barrett Browning used it in her book-length poem *Aurora Leigh* (1856).
Romney, heir to the Leigh fortunes, spots his cousin Aurora one summer morning in the garden where he plans to make a marriage proposal. Aurora is pretending "in sport, not pride" to crown herself as
a poet on the day of her twentieth birthday. Romney watches as she
decides which plant to use for a crown. Ruling out bay, verbena,
guelder-rose, and apple blossoms, she decides,

> Ah—there's my choice,—that ivy on the wall,
> That headlong ivy! not a leaf will grow

But thinking of a wreath. Large leaves, smooth leaves,
Serrated like my vines, and half as green.
I like such ivy, bold to leap a height
'Twas strong to climb; as good to grow on graves.
As twist about a thyrsus; pretty too
(And that's not ill) when twisted round a comb.

In "When the Hounds of Spring are on Winter Traces," a chorus
from "Atalanta in Calydon" (1865), Algernon Charles Swinburne described the poet's love interest, a "maiden most perfect, lady of light":

The ivy falls with the Bacchanal's hair
Over her eyebrows shading her eyes;
The wild vine slipping down leaves bare
Her bright breast shortening into sighs;
The wild vine slips with the weight of its leaves,
But the berried ivy catches and cleaves
To the limbs that glitter, the feet that scare
The wolf that follows, the fawn that flies.

In "Crimean Sketches" (ca. 1857), translated by Dorian Rottenberg,
Alexei Tolstoy compared the strength of the ivy with the strength of
love:

But now young ivy, twining round
Its walls, war's traces has concealed.
Has my love not yet likewise wound
Around your wounds, have they not yet healed?

The ivy is perhaps the most neglected of plants in a garden because
it is not always purposely cultivated. Yet it is tenacious, quietly covering and hiding ruins beneath. A popular misconception is that ivy
chokes its victims—stone walls, trees, buildings, and other plants—
and derives its full nutrition from them rather than its own root system. Ivy may stunt the growth of some trees on which it grows, but

it is not a parasitic plant, contrary to the thought expressed in a poem by Charles Dickens, called "The Ivy Green" from *Pickwick Papers* (1837):

> For the stateliest building man can raise,
> Is the Ivy's food at last.
> Creeping on, where time has been,
> A rare old plant is the Ivy green.

JASMINE

Throwing wide her elegant sweets

Both *Jasminum* in the Oleaceae family and *Gelsemium* in the Loganiaceae are garden plants commonly called jasmine or jessamine. *Jasminum* consists of about 200 species of woody climbers and shrubs mostly native to Europe and Asia. Most have fragrant flowers, which are the source of jasmine perfumes, colognes, and teas. The Italian yellow jasmine, *J. humile*, is an evergreen native from the Middle East to Southeast Asia. The winter jasmine, *J. nudiflorum*, is a deciduous native of northern China; its flowers appear in late winter before the leaves emerge. The common or true jasmine from Asia Minor, *J. officinale*, has been in cultivation for hundreds of years.

Gelsemium is a genus of three species, two of which are North American natives ranging from Central America to the southeastern United States. *Gelsemium sempervirens* is a woody evergreen climber commonly called Carolina or yellow jessamine; it has fragrant tubular flowers that make a showy appearance in early spring as the vines sprawl upward and over small-to-medium-sized trees and shrubs that have not yet produced their leaves. *Gelsemium rankinii* is little known outside the American Southeast, but it bears a profusion of odorless flowers in the fall and in the spring. To distinguish them from the true jasminums, they are sometimes given the name false jasmine.

Literary references apply generally to *Jasminum*, which receives its name from a medieval Latinization of the Persian name for the plant, *yasmin* or *yasamin*. *Gelsemium* is the Latinized spelling of the Italian name for the jasmine, *gelsomino*.

In Theodore T. Barker's translation of Léo Delibes's opera *Lakmé* (1883), set in British-occupied India, the title character and her servant Mallika while away their time in a sheltered canopy of jasmine:

> 'Neath the dome, the jasmine
> To the roses comes
> 'Neath the leafy dome
> Where the jasmine white
> To the roses comes greeting
> By flower banks fresh and bright greeting.

Jasmine appeared in the early poetry of Hindu India as the "Moonlight of India." It was often referred to as a flower of the night because its fragrance permeates the evening air. Thomas Moore evoked such a perfumed evening in "Jasmine":

> Many a perfume breathed
> From plants that wake when others sleep;
> From timid jasmine-buds that keep
> Their odour to themselves all day,
> But when the sunlight dies away
> Let the delicious secret out
> To every breeze that roams about.

Poets have been moved to proclaim the beauty of the jasmine. William Cowper, a pre-romantic English poet, described the enchanting flowers in "The Winter Walk at Noon" from *The Task*, a contemplation upon God, nature, and humanity:

The jasmine throwing wide her elegant sweets,
The deep dark green of whose unvarnished leaf
Makes more conspicuous and illumines more,
The bright profusion of her scattered stars.

English poet and clergyman Charles Churchill, known for his satire, wrote with admiration of the exquisitely simple flower in "The Perfume of Jasmine":

The jasmine, with which the queen of flowers,
To charm her god, adorns his favourite bowers;
Which brides, by the plain hand of neatness drest,—
Unenvied rival!—wear upon the breast.
Sweet as the incense of the morn, and chaste
As the pure zone which circles Dian's waist.

A legend associated with jasmine, which may be based in fact, involves Pope Sixtus III, who dreamed that a great quantity of snow would fall in the month of August. When his dream came true, he built the Borghese Chapel in Rome to commemorate the divine message he had received. He decreed that on the anniversary of the snow day, a recreation of the snow shower should be staged in the chapel during services. The fragrant flowers of white jasmine were used for snow. It is said that the fragrance of the flowers was so powerful that some would-be worshippers were deterred from attending the ceremony.

In his published journals, English naturalist Gilbert White also noted the jasmine's scent: "July 17 [1783] The jasmine, now covered with bloom, is very beautiful. The jasmine is so sweet that I am obliged to quit my chamber." English journal editor and poet Thomas Hood wrote in "Flowers" that "Jasmine is sweet and has many loves."

LAUREL & DAPHNE

Oh, who can love the laurel wreath

The fifty species of *Daphne* in the family Thymelaeaceae are Eurasian in distribution, most being small evergreen or deciduous shrubs with fragrant flowers. Well-known daphnes are the winter-blooming *D. mezereum*; the variegated *D.* ×*burkwoodii* 'Carol Mackie'; the charming dwarf Hungarian *D. arbuscula*; the sweet-smelling *D. odora*; and *D. cneorum*, the garland flower.

The word daphne is the Greek name for *Laurus*, the laurel or bay tree; Linnaeus chose to name the unrelated genus *Daphne* because many of the species resemble the true laurel of the family Lauraceae, *Laurus nobilis*, which is also the culinary bay. The original crown of laurelsis generally believed to be made from the true laurel. Some species of both the laurel and daphne grow to-gether in the Mediterranean area. Other plants with the common name of lau-rel or bay include the Alexandrian lau-rel or poet's laurel, *Danaë racemosa*, in the family Liliaceae, and the New World sweet bay, *Magnolia virginiana*. In early literature, the daphne, bay, laurel, and bay laurel are generally interchanged, and it is not always apparent which spe-cific plant is meant.

In an old Romanian story, a young nymph known as the Daughter of the Laurel dwelt among the branches of a lau-rel bush. Each evening the laurel opened its branches so the young nymph could go to her nightly dance in the Valley of the Flowers. One night during her pirouettes, a handsome youth saw her and praised her beauty. He expressed his passion for her and started to embrace her. Because of her shyness, the laurel nymph fled and disappeared into the flowery groves where she was protected by the Star Queen, who lived in a palace in the clouds. Undaunted, the handsome stranger searched for and found the laurel nymph and sang her to sleep in his arms, promising never to abandon her. When she awoke, he was gone. The nymph cried to the Star Queen to find him for her. But the Queen did not respond to her pleas, and the laurel tree, too, refused to open its branches for her to return as daylight approached. The laurel told her, "The wreath of honor has fallen from thy brow; there is no longer any place for you here." As the sun rose over a nearby mountain and shone on the Daughter of the Laurel, she dissolved away into the morning dew.

The roots of this story come from the Greek myth of the origin of the laurel. Daphne was a maiden nymph who was amorously pursued by Apollo (called Phoebus by the Romans), whom she distrusted. Diana answered Daphne's pleas for help by changing her into a laurel to save her from his solicitations. When Apollo started to embrace Daphne and realized he was kissing a tree with a crown of green leaves, he decreed that the tree would forever be sacred. In 1621 English writer George Sandy translated Ovid's *Metamorphoses*, including the description of Daphne's change from nymph to shrub:

> And slender filmes her softer sides invest,
> Haire into leaves, her armes to branches grow:
> And late swift feet, now rootes, are less than slow.
> Her gracefull head a leavy top sustaynes,
> One beauty throughout all her forme remaines.
> Still Phoebus loves. He handles the new plant;
> And feeles her heart within the barke to pante.

In Richard Strauss's one-act opera *Daphne*, the story takes an interesting twist. Here, Daphne is a childlike maiden who loves best to spend her days on the hillsides communing with nature. Her child-hood playmate Leukippos has grown into a shepherd whose thoughts are turning toward amorous adult pursuits. Spurred on by the carnal atmosphere of the approaching feast of Dionysus, Leukippos makes advances toward Daphne but is rejected. And Daphne's mother, en-couraging her daughter to turn away from childish playing on the hill-sides, offers jewels and elegant clothing for Daphne to wear to the feast. The girl refuses them, however, favoring her simple country clothes. Leukippos then takes the new clothes, thinking to wear them to the feast to trick Daphne into trusting him, thereby winning her love. Mean-while, Apollo is intrigued by the worldly preparations for the feast of Dionysus and comes down from Mt. Olympus disguised as a shep-herd. He sees Daphne, is smitten, and pursues her. At first she is at-tracted to the god she believes is a shepherd, but the jealous Leukippos unmasks Apollo, shocking Daphne with his identity as the god. Amid thunder and rhapsodic outbursts of rage, Apollo kills Leukippos. But Daphne throws herself onto the body of her childhood friend. Apol-lo sees her sincerity and transforms Daphne into a laurel tree, vowing that branches cut from her tree will be worn by the bravest of mortals.

Several poems mention Daphne and Apollo. For example, Andrew Marvell in "The Garden" (1667) explained:

> When we have run our passion's heat,
> Love hither makes his best retreat.
> The gods, that mortal beauty chase,
> Still in a tree did end their race.
> Apollo hunted Daphne so
> Only that she might laurel grow.
> And Pan did after Syrinx speed,
> Not as a nymph, but for a reed.

Marvell's poem compares the quiet, peace, and beauty in a garden to the trouble and stress outside it. In "The Laurel" Wordsworth also re-told the Daphne and Apollo story:

> The leaves of any pleasant tree
> Around the golden hair,
> Till Daphne, desperate with pursuit
> Of his imperious love
> At her own prayer transformed, took root
> A laurel in the grove.

Despite Apollo's lack of success in his pursuit of Daphne, the crown or garland of laurel became emblematic of victory for both the Greeks and Romans; no glory was greater than to be honored by a laurel crown. The Romans gave wreaths of it to the winners of the Pythian games, but the victors of the Greek Olympic games received a wreath of wild olive leaves. In the language of flowers, laurel signified glory, which symbolism has survived in the honors of Nobel Laureate and Poet Laureate. The Romans decorated with laurel for it communicated poetry, rhetoric, and prophecy; for that reason people put its leaves under their pillows for inspiration. The tree was considered a safeguard against lightning. And the death of a laurel or bay tree was an omen of impending disaster.

Chaucer, writing in 1381 in the *Parliament of Fowls*, made the earliest use of laurel in English literature with "The victor palm, laurer [laurel] to deuyne." And daphne made its first appearance in 1430 in English poet and priest John Lydgate's *Complaint of the Blacke Knight*: "I sawe the Daphene closed under rynde, Grene laurer and the holsome pyne." But even earlier, ca. 300 BCE, the Greek epic poet Anyte, who has been called the female Homer, wrote in "Under a Laurel" (translated by R. A. Furness),

> Sit all beneath fair leaves of spreading bay
> And draw sweet water from a timely spring,
> And let your breathless limbs, this summer day,
> Rest, in the west wind's airy buffeting.

William Habington, best known for his collections of love poems, wrote *Castara* (1634), a collection that celebrates his wife, Elizabeth Herbert: "No name can tame/The North-wind's fury, but Castara's name/Climbe yonder forked hill, and see if there barke of every Daphne, not appeare Castara written." The poem is "My Honoured Friend, Mr. E. P.," a dedication to Endymion Porter, member of parliament and groom of the bedchamber to Charles I. The poem invokes Castara as poet, muse, and "Deity of her sex."

The laurel has long figured in the decorating of the graves of fallen heroes and loved ones. Milton's elegiac poem "Lycidas" (1637), lamenting the drowning of a classmate in the Irish Sea, refers to the laurels that will be placed on the grave:

> Yet once more, O ye Laurels, and once more
> Ye myrtles brown, with Ivy never sear,
> I come to pluck your Berries harsh and crude,
> And with forc'd fingers rude,
> Shatter your leaves before the mellowing year.
> Bitter constraint, and sad occasion dear,
> Compels me to disturb your season due:
> For Lycidas is dead, dead ere his prime
> Young Lycidas, and hath not left his peer.

The fallen soldiers of the American Civil War were given poetic laurels. In "The March into Virginia" Herman Melville wrote of the brave mortals of the first battle of Manassas (21 July 1861), also called Bull Run:

> All they feel is this: 'tis glory,
> A rapture sharp, though transitory,
> Yet lasting in belaureled story.
> So they gayly go to fight,
> Chatting left and laughing right.

Likewise, Henry Timrod used laurel as a symbol of brave mortality in an ode on the occasion of decorating the Confederate graves in Magnolia Cemetery in Charleston, South Carolina, June 1866:

> In seeds of laurel in the earth
> The blossom of your fame is blown,
> And somewhere, waiting for its birth,
> The shaft is in the stone!

On the other hand, English poet Eliza Cook wrote "The Wreaths" in which she shunned the laurel because of its association with the dead:

> Whom do we crown with the laurel-leaf?
> The hero-god, the soldier chief
>
> . . . .
>
> Oh, who can love the laurel wreath,
> Plucked from the gory field of death?

LILAC

Oh, were my love yon lilac fair

The common lilac of gardens, *Syringa vulgaris,* is one of more than twenty deciduous species of shrubs and trees of the Oleaceae, the olive family, that are distributed from southeastern Europe to eastern Asia. Several species are grown as ornamentals mainly for their showy panicles of highly fragrant flowers. Popular ornamental cultivars include *S. ×hyacinthiflora; S. reticulata,* the Japanese tree lilac with creamy white flowers; *S. ×laciniata,* a small shrub with prettily dissected leaves; and *S. ×persica,* the Persian lilac. Lilacs have been known in England since the time of Henry VIII in the early sixteenth century, having made their way through Vienna from the Middle East.

The name lilac is from an old Persian word *nilak* and Arabic *lilah,* for blue or bluish, referring to the color of its flowers. The name lilac was sometimes also applied to mean flowers in general. In English writings it also appeared as late as the nineteenth century as laylock and lilach. The Latin name *Syringa* is from the Greek word *syrinx,* for pipe, in reference to the plant's hollow stems—thus the origin of its older common name, the pipe bush. Linnaeus took for the garden lilac the name *Syringa* from the sweet-smelling mock orange which the Greeks had called syringa. He then gave to the mock orange the name *Philadelphus,* after Ptolemy Philadelphus, an Egyptian king reigning at about

300 BCE. The reasoning behind this exchange of names has eluded botanists ever since. As a common name, Syringa continues to refer to *Philadelphus* in some references.

According to Greek mythology, Pan, the god of pastures and forests, formed a musical pipe from the hollow stems of the lilac or some other plant. Pan tried to woo Syrinx, an Arcadian mountain nymph, but desiring to remain chaste, Syrinx fled from Pan through the forests to the banks of the River Ladon. Pan leapt to catch her, but the water nymphs came to her rescue by transforming her into a stand of reeds. A gentle whistling sound arose as the wind blew through the reeds, inspiring Pan to cut them and bind them together into the panpipe or syrinx. Thus began the instrument's fabled use by shepherds. The story was told by the Roman poet Ovid in *Metamorphoses* and was translated into English in part by poet and critic John Dryden:

> He filled his arms with Reeds, new rising on the place;
> The tender canes were shaken by the wind,
> And breathed a mournful air, unheard before;
> That much surprising Pan, yet pleased him more.
> Admiring this new music, 'Thou,' he said,
> 'Who can'st not be the partner of my bed,
> At least shall be the consort of my mind;
> And often, often to my lips be joined.'
> He formed the Reeds, proportioned as they are,
> Unequaled in their length, and waxed with care,
> They still retain the name of his ungrateful fair.

The story was told with a different twist by Andrew Marvell in "The Garden" (1667). He saw a higher goal behind Pan's attraction for Syrinx:

> Apollo hunted Daphne so
> Only that she might laurel grow;

And Pan did after Syrinx speed
Not as a nymph, but for a reed.

Lilacs appeared in novels principally during the 1800s, but often as passing references. Their wide and frequent use suggests that lilacs were well known and loved. For example, Emily Brontë's *Wuthering Heights* (1847) conjures only the color of lilacs: "There's a little flower up yonder—the last bud from the multitude of bluebells that clouded those turf steps in July with a lilac mist." In the conclusion to *Silas Marner* (1861) George Eliot, whose real identity was Mary Ann Evans, wrote, "There was one time of the year which was held in Raveloe to be especially suitable for a wedding. It was when the great lilacs and laburnums in the old-fashioned gardens showed their golden and purple wealth above the lichen-tinted walls, and when there were young calves still young enough to want bucketfuls of fragrant milk."

In *The Return of the Native* (1878), Thomas Hardy suggested lilac for the particular color of clouds at dusk: "The sun, resting on the horizon line, streamed across the ground from between copper-coloured and lilac clouds, stretched out in flats beneath a sky of pale soft green." Edith Wharton wrote in the short story "Mrs. Manstey's View" (1891) of both the lilac color and the plant: "In the very next enclosure did not a magnolia open its hard white flowers against the watery blue of April? And was there not, a little way down the line, a fence foamed over every May by lilac waves of wistaria? Farther still, a horse-chestnut lifted its candelabra of buff and pink blossoms above broad fans of foliage; while in the opposite yard June was sweet with the breath of a neglected syringa, which persisted in growing in spite of the countless obstacles opposed to its welfare." Wharton's scene recalls many a neighborhood lilac, neglected but persistently beautiful.

In poetry, Robert Burns wrote of lilacs:

Oh, were my love yon lilac fair
Wi' purple blossoms to the spring,

And I a bird to shelter there,
When wearied on my little wing.

And English satirist Thomas Hood recalled the lilacs of his child-hood in "The Plea of the Mid-summer Fairies" (1827):

> I remember, I remember.
> The roses, red and white,
> The violets, and the lily-cups—
> Those flowers made of light!
> The lilacs, where the robin built,
> And where my brother set
> The laburnum on his birthday—
> The tree is living yet!

Rupert Brooke, a handsome young English poet and darling of literary circles prior to World War I, wrote in "The Old Vicarage, Grant-chester" (1912),

> Just now the lilac is in bloom,
> All before my little room;
> And in my flower-beds, I think,
> Smile the carnation, and the pink.

English poet, critic, and antimodernist Alfred Noyes giddily wrote in 1908 in "The Barrel Organ":

> Go down to Kew in lilac-time, in lilac-time, in lilac-time;
> Go down to Kew in lilac-time (it isn't far from London!)
> And you shall wander hand in hand with love in summer's
>   wonderland;
> Go down to Kew in lilac-time (it isn't far from London!).

However, probably the most famous mention of lilacs in American literature is by poet and journalist Walt Whitman, who wrote one of his wartime poems, "When Lilacs Last in the Dooryard Bloom'd" (1865), to mourn the death of Abraham Lincoln:

> When lilacs last in the dooryard bloom'd,
> And the great star early droop'd in the western sky in
>   the night,
> I mourn'd, and yet shall mourn with ever-returning
>   spring.
> Ever-returning spring, trinity sure to me you bring,
> Lilac blooming perennial and drooping star in the west,
> And thought of him I love.

The origin and early cultivation of lilacs in the gardens of Persia was pondered by Lydia Huntley Sigourney, the "Sweet Singer of Hartford," who was one of the first American women to enjoy a successful literary career, publishing some twenty-two volumes of poetry in the 1800s:

> Lilac of Persia! Tell us some fine tale
> Of Eastern lands; we're fond of travellers.
> Have you no legends of some sultan proud,
> Or old fire-worshipper? What! not one note
> Made on your voyage? Well, 'tis wondrous strange
> That you should let so rare a chance pass by,
> While those who never journeyed half so far
> Fill sundry volumes, and expect the world
> To reverently peruse and magnify
> What it well knew before!

In the language of flowers, the white lilac symbolized purity, modesty, and youthful innocence, and the purple lilac suggested the first emotions of love: "The fragrant lilac with its soft colors evokes strong memories that time does not easily erase." The following "Persian Love-

Song" about first love suggests that the lilac is the most suitable flower to symbolize lasting love:

> Ah, let me weave a chaplet for your hair,
> Of pale and rosy lilacs, lady fair.
> Woe to the lover who would choose a rose
> That in its heart a stinging bee may close.
> Or yet a lily, or a spray of vine,
> Or any bloom that wreathes a cup of wine.
> The flower I gather, love, for your sweet sake
> Breathes love that neither time nor ill can shake.

LILY

To feed in the garden and to gather lilies

The family of lilies consists of at least 3000 species within 250 genera, ranging from *Agapanthus* to *Zigadenus*. The Liliaceae is now split into smaller families that more accurately reflect relationships. About 100 species are bulbous plants of the genus *Lilium*, which are native primarily to temperate areas of the Northern Hemisphere. Whereas true lilies are in the genus *Lilium*, numerous other plants are commonly called lilies, such as canna, atamasco, belladonna, blackberry, Guernsey, and zephyr. Some familiar true lilies are *L. canadense*, the Canada lily; *L. candidum*, the Madonna lily; *L. martagon*, the Turk's cap lily; *L. superbum* of the eastern United States; *L. regale*; and their numerous hybrids and cultivars.

The Latin name *Lilium* is from the Greek *leirion*, the name applied to the eastern Mediterranean–native Madonna lily (now *L. candidum*), one of the oldest known lilies. The specific epithet *candidum* means shining or pure white. It has a strong claim to being the oldest domesticated flower, perhaps because of its simple loveliness. Appropriately, in the language of flowers, the lily symbolized majesty and stateliness. In the church, it has come to mean purity, particularly for weddings, while also serving as the main floral symbol of the Virgin Mary. The Madonna lily was venerated as a sacred flower because its petals suggested a spotless body and its golden anthers a soul gleaming with heavenly light. The Madonna lily is used as a decoration on various church days associated with the Virgin Mary. The white lilies, with their suggestions of purity and chastity, are associated also with certain male and female saints.

Lilium and some of its allied genera have been cultivated since at least ca. 2800 BCE. In early Cretan civilization lilies (probably *L. candidum*) were molded onto gold jewelry and pottery. The brightly colored *L. chalcedonicum*, the scarlet Turk's cap, is depicted on frescoes unearthed in Greece that date to 2000 BCE. The Madonna lily appeared frequently in art, particularly in depictions of the Virgin Mary in the fifteenth century. European and North American gardens have included their native species for some time, but the Asiatic species (*L. speciosum* and *L. lancifolium*) did not reach the West until the nineteenth century.

In Greek mythology the Milky Way and the lily have a common origin. Zeus wanted his illegitimate son Heracles, or Hercules, to be made an infant god and immortal, but Hera, Juno to the Romans, Zeus's wife and stepmother of Heracles, objected strongly. Zeus ordered Somnus, the god of sleep, to give Hera a draught to make her sleep. While she was in a deep slumber, Zeus put the child to Hera's breast so that Heracles would drink her milk and become immortal. The hungry child sucked so greedily from Hera's breasts that milk flowed rapidly and splashed across the heavens to form the Milky Way. The drops that fell to earth became white lilies.

The King James Bible contains at least fourteen citations of lilies, although the true plant mentioned is not known. *Lilium candidum* and *L. chalcedonicum* are native to the Middle East and could be the lilies to which biblical authors referred; the rose of Sharon in the Song of Solomon (2:1) is presumed to be one of those species. Other references include:

> Consider the lilies how they grow: they toil not, they spin

not; and yet I say unto you, that Solomon in all his glory
was not arrayed like one of these. (Luke 12:27)

And the chapiters that were upon the top of the pillars were
of lily work in the porch, four cubits. (1 Kings 7:19, and
similarly 7:22)

I will be like the dew to Israel; he will blossom like a lily.
Like a cedar of Lebanon he will send down his roots. (Hos.
14:5)

Some suggest that lily in Hebrew (Shusan, the origin of the name
Susan) referred to any beautiful flower. Translators of the King James
Bible must have had difficulty in finding the right plant name for use
in the erotic poetry of the Song of Solomon:

Thy two breasts are like two young roes that are twins,
which feed among the lilies. (Song of Sol. 4:5)

My beloved is gone down into his garden, to the beds of
spices, to feed in the garden and to gather lilies. I am my
beloved's, and my beloved is mine: he feedeth among the
lilies. (Song of Sol. 6:2–3)

Thy navel is like a round goblet, which wanteth not liquor:
thy belly is like a heap of wheat set about with lilies. (Song
of Sol. 7:2)

And in the "Battle Hymn of the Republic," Julia Ward Howe echoes
the religious connotations with "In the beauty of the lilies, Christ was
born across the sea."

Shakespeare also was not always clear in his references to lilies. In
act 3 of *Troilus and Cressida* (1602), Troilus requests "swift transportation
to those fields/Where I may wallow in the lily beds." In act 3 of *Henry
VIII* (1613), Queen Katherine tells Cardinal Wolsey, "Almost no grace
allow me: like the lily,/That once was mistress of the field and flour-

ish'd/I'll hang my head and perish." In *The Life and Death of King John* (1597)—a historical play dealing with the events of King John's reign, excluding those relative to the Magna Carta—the Earl of Salisbury speaks to the king in the state room of the palace, discussing their demands in exchange for surrendering the crown:

> Therefore, to be possess'd with double pomp,
> To guard a title that was rich before,
> To gild refined gold, to paint the lily,
> To throw a perfume on the violet,
> To smooth the ice, or add another hue
> Unto the rainbow, or with taper-light
> To seek the beauteous eye of heaven to garnish,
> Is wasteful and ridiculous excess.

The perfection of the lily moved Englishman Ben Jonson in "The Turne" from *The Underwood* to note that life may be perfect in short measures, just as the lily is:

> A lily of a day
> Is fairer far, in May,
> Although it fall and die that night;
> It was the plant and flower of light.

In "The Lilly," William Blake described the serenity of the plant by comparing it to the barbs and threats of others:

> The modest rose puts forth a thorn,
> The humble sheep a threat'ing horn,
> While the lilly white shall in love delight,
> Nor a thorn, nor a threat, stain her beauty bright.

Other writers made reference to the lily's pure white. Walafrid Strabo's knowledge of the lily—he was a monk at the Benedictine abbey on Reichenau, a small island in Lake Constance (Bodensee) in Germany—suggests that the flower reached Middle Europe before the first Crusades of the late 1000s. Strabo observed in his *Hortulus* that neither the whiteness of Parian marble, a singular type of stone from the Aegean island of Paros, nor the fragrance of the spikenard (*Nardostachys*) can surpass the lily, which glistens as white as snow. Chaucer wrote of a *cote-armour*, a covering of rich material knights wore over their armor, "as whyt as is a lily-flour" in *The Tale of Sir Thopas* from *The Canterbury Tales* (1387–1400). Leigh Hunt compared an angel to a lily in "Abou Ben Adhem and the Angel" (1838):

> Abou Ben Adhem (may his tribe increase!)
> Awoke one night from a deep dream of peace,
> And saw, within the moonlight in his room,
> Making it rich, and like a lily in bloom,
> An angel writing in a book of gold.

Abou Ben Adhem goes on to tell the angel that he loves his fellow man, for which he receives God's blessing. Hunt wrote poetry and drama and founded several journals, including the *Examiner*, for which he was imprisoned for two years on charges of libel against the prince regent, later King George IV.

An unusual variation on the poetic use of the lily's whiteness comes in "The Nymph Complaining for the Death of her Fawn." The English metaphysical poet Andrew Marvell compared the coloration of a fawn to the lily:

> Among the beds of lilies, I
> Have sought it oft, where it should lie,
> Yet could not, till itself would rise,
> Find it, although before mine eyes;
> For, in the flaxen lilies' shade,
> It like a bank of lilies laid.

Tennyson made a similar comparison in "Lady Clare" (1842):

> It was the time when lilies blow,

> And clouds are highest up in air,
> Lord Ronald brought a lily-white doe
> To give his cousin, Lady Clare.

The purity of the lily has understandably been used by poets to give paeans to their objects of love. Thomas Campion (or Campian) in the *Fourth Book of Airs* described the beauty of his lady love in this untitled poem from 1617:

> There is a garden in her face,
> Where roses and white lilies grow;
> A heavenly paradise is that place
> Wherein all pleasant fruits do flow.

In "The Rhine" from canto 2 of *Childe Harold's Pilgrimage* (1812) by Lord Byron, Harold wants to catch the eye of his intended with flowers gathered on the banks of the Rhine:

> I send the lilies given to me;
> Though long before thy hand they touch
> I know that they must wither'd be,
> But yet reject them not as such.

It is generally assumed that the travels and reflections of Harold are in fact Byron's, which Byron denied. Almost certainly the lilies refer to the scandalous reports widely circulated about an incestuous relationship between Byron and Augusta Leigh, his half sister.

In "The Blessed Damozel" (1850), Dante Gabriel Rossetti described a woman with sacramental symbolism: three lilies in her hand, seven stars in her hair, and a white rose on her robe. The woman leans out from heaven to watch the world below:

> And still she bowed herself and stopped
> Out of the circling charm;
> Until her bosom must have made

The bar she leaned on warm,
And the lilies lay as if asleep
Along her bended arm.

And Gilbert and Sullivan wrote a deliberate operatic satire of the Aesthetic Movement of the late nineteenth century called *Patience* (1881), in which Oscar Wilde is mocked through the character Reginald Bunthorne, "a fleshly poet," who carries a lily and sings, "Let me confess! / A languid love for lilies does not blight me!"

LILY OF THE VALLEY

Emblem of a lovelier thing

Only four species of lily of the valley grace the family Liliaceae (or the smaller, more recently defined Convallariaceae). *Convallaria majalis*, the one horticulturally significant species, is native to temperate regions of the Northern Hemisphere. It has become naturalized in North America. It typically flowers in late spring on an arching scape bearing fragrant petals that are white. Several selections exist, including 'Albostriata', with white-striped leaves; 'Flore Pleno', with double flowers; and var. *rosea*, with pink flowers.

The Latin word *convallis* means valley, and *majalis* means of the month of May, suggesting the flowering period. The common name, lily of the valley, is a literal translation from the old medieval name for the plant, *Lilium convallium*. Because of its blooming time, it is associated with Whitsunday, or Pentecost, the seventh Sunday after Easter in the Christian church year, which celebrates the Holy Spirit descending to the apostles. Observed around 15 May, Whitsunday calls for the recently baptized to wear new white clothes. Secular traditions of mid-May claimed the lily of the valley as well. Renters in Hessian townships had to pay a bouquet of the mayflowers as rent by that date. The fifteenth of May is also the traditional day under old Scottish law for the removal of rural tenants who had not paid their rent.

One legend associated with the lily of the valley comes from Sussex, England. The hermit St. Leonard battled a dragon that was the devil in disguise. He struggled for four days and finally summoned the strength to cut off the dragon's head, not before, however, his adversary's strong claws had torn through his armor and spilled his blood. Lilies of the valley sprang up from the drops of blood, and pilgrims to the site could trace the path through the fields and woods where the battle had raged. Every year since, lilies of the valley bloom each May on the battleground.

The Cherokee Indians of eastern North America have a legend that tells of Little Dawn Bird who was taught plant lore by her father Big Tree. One day, she was to go alone to the mountains, but her father was worried for her safety and followed secretly behind her. Thinking she was alone, she marked her trail by dropping white quartz pebbles on leaf litter along the way. When she grew tired and fell asleep, Big Tree quietly watched over his daughter from nearby. Suddenly, the girl awoke as a wildcat pounced upon her. But the instant before, the vigilant Big Tree had pierced the wildcat's heart with an arrow as it began its leap. Still not wanting to reveal his presence, Big Tree watched as Little Dawn Bird regained her composure and searched for her way back home. Looking for the little pebbles she had placed to mark her trail, she found they had turned into the tiny white bells of lily of the valley. Their tinkling music guided Little Dawn Bird safely home to her waiting father. Since lily of the valley is not native to North America, either this Cherokee legend may be of recent origin or the plant name may have changed through the years.

John Gerard called *Convallaria* the May lily, a name that survives in some regions of England, and recommended it for "griefe of the gout." Lily of the valley used to be known as glovewort because it cured sores on the hands. Likewise, witches recommended rubbing its blossoms on the face to cure freckles. The plant was also known as Our Lady's tears, a reference to the Virgin Mary, to whom it is dedicated. In the language of flowers, lily of the valley signified the return of happiness, because the blooms are the harbinger of the warm days of spring.

Robert Louis Stevenson was an Edinburgh-born writer who enjoyed a Bohemian lifestyle while producing a vast canon of writings. His novel *Kidnapped* (1886) tells of the young David Balfour who is kidnapped by his miserly uncle Ebenezer to keep control of the Balfour estate. In chapter 1, as the "pretty lad" David leaves his dead parents' manse for the last time, Mr. Campbell, the minister of Essendean, gives him a packet of material left for David by his father. Among the items is a "little piece of coarse yellow paper, written upon thus in red ink":

> To make Lilly of the Valley Water—Take the flowers
> of lilly of the valley and distil them in sack, and drink
> a spooneful or two as there is occasion. It restores speech
> to those that have the dumb palsey. It is good against the
> Gout; it comforts the heart and strengthens the memory;
> and the flowers, put into a Glasse, close stopt, and set into
> ane hill of ants for a month, then take it out, and you will
> find a liquor which comes from the flowers, which keep in
> a vial; it is good, ill or well, and whether man or woman.
> And then in the minister Campbell's hand was appended:
> 'Likewise for sprains, rub it in; and for the cholic, a great
> spooneful in the hour.'

Many poets have been inspired by the purity and simplicity of the lily of the valley. In "The Lily of the Valley," Bishop Richard Mant wrote,

> Fair flower, that lapt in lowly glade,
> Dost hide beneath the greenwood shade
> Than whom the vernal glade
> None fairer wakes, on bank, or spray,
> Our England's lily of the May.
> Our lily of the vale!

In the early nineteenth century Leticia Elizabeth Landon wrote numerous volumes of poetry under the initials L. E. L. She endured various vague scandals of adultery and died from an overdose of prussic acid, but not before she had composed the beautiful lines of "A History of the Lyre" about the beautiful Eulalie, an orphaned lute player of ancient Rome:

> Look at those flowers near yon acacia tree—
> The lily of the valley—mark how pure
> The snowy blossoms,—and how soft a breath
> Is almost hidden by the large dark leaves.
> Not only have those delicate flowers a gift
> Of sweetness and of beauty, but the root—
> A healing power dwells there; fragrant and fair
> But dwelling still in some beloved shade.

William Cullen Bryant compared the two perfect forms of a lily and a child:

> Innocent child and snow-white flower!
> Well are ye paired in your opening hour:
> Thus should the pure and the lovely meet,
> Stainless with stainless, and sweet with sweet.

Scottish poet James Thomson wrote a collection of four poems called *The Seasons*. In "Spring" (1728) he described the influence of the season on nuptial love and nature, including, "Where scatter'd wild the lily of the vale/Its balmy essence breathes." William Wordsworth, in "The Excursion" (1814), took note of "That shy plant . . . the lily of the vale,/That loves the ground."

George Croly, rector of St. Stephen's Walbrook, who Byron in *Don Juan* called Rev. Rowley Powley, published numerous narrative and romantic poems, including the tribute "The Lily of the Valley":

> White bud! that in meek beauty so dost lean
> The cloistered cheek, as pale as moonlight snow,
> Thou seem'st beneath thy huge high leaf of green

An Eremite beneath his mountain's brow.
White bud! thou'rt emblem of a lovelier thing
The broken spirit that its anguish bears
To silent shades, and there sits offering
To Heaven the holy fragrance of its tears.

LYCHNIS

A garden of pleasure

Lychnis is a member of the Caryophyllaceae or pink family, and its twenty species are widely distributed in the Northern Hemisphere. Campion, rose campion, and catchfly are common names shared with the closely related genus *Silene*. Some taxonomists have even merged the two genera. Common names applied only to *Lychnis* are lamp flower and ragged robin. Species include *L. coronaria*, which has densely woolly leaves; *L. chalcedonica*, the Maltese cross, with hairy leaves; and *L. flos-jovis* of the Alps.

    The lychnis or rose campion obtains its Latin name from the Greek word *lychnos*, which means lamp, because the hairy stout leaves of *Lychnis coronaria* were formerly used as lamp wicks, and because the flowers glow like flames. In the church, *Lychnis* is dedicated to St. John the Baptist. Since *L. coronaria* blooms at the time of his feast day, 24 June, the flame-colored flowers are like a "light to them which sit in darkness." An older Latin name for the plant was *Candelabrum*, which means candle holder. The common name campion may have been derived from the word champion, because of its use in victory garlands at sporting events. The specific epithet *coronaria* means crown, another reference to the garlands of winners; however, the name may derive from *campagne*, the French word for country, giving the plant name the meaning "flower of the fields." The name of *L. chalcedonica* means that it is from Chalcedon (Kadikoy) in Turkey near Istanbul. It is called the Maltese cross because the plant was introduced to northern European gardens during the Crusades, or more likely because the shape of the flower resembles the Cross of Malta. The species name *flos-jovis* literally means "flower of Jupiter." The common name catchfly carried its significance into the nineteenth century, for in the language of flowers, it suggested a snare, youthful love, and betrayal. The name catchfly applies to those species of both *Lychnis* and *Silene* that have sticky and viscid stems that trap small insects.

Among the earliest written English references to lychnis is in Philemon Holland's translation of *Pliny's Natural Historie* (1601): "As touching Lychnis, that flaming hearbe surnamed Flammea." At about the same time, John Gerard in *The Herbal or General History of Plants* (1633 revision by Thomas Johnson) described using catchfly: "I have called it Catchflie, or Lime woort . . . The whole plant, as well leaues and stalkes, as also the floures, are here and there couered ouer with a most thicke and clammie matter like unto Birde lime."

John Parkinson provided simple but elegant praise of the campion in his *Paradisi in Sole, Paradisus Terrestris* (1629): "There be divers sorts of Campions, as tame as wilde, and although some of them that I shall here entreate of, may peradventure be found wilde in our owne countrye, yet in regarde of their beautifull flowers, they are to be respected, and noursed up with the rest, to furnish a garden of pleasure; as for the wild kindes, I will leave them for another discourse."

In Samuel Taylor Coleridge's "Picture," the vision of a fine lady includes "The heads of tall flowers that behind her grow/Lychnis, and willow-herb, and foxglove bells."

John Clare roamed the early nineteenth-century woods and hills of his native Northamptonshire and noticed "ragged-robins by the spinney lake" in his poem "Wild Flower Nosegay." And in "Noon" (1809), writing of a warm summertime, he described

> Ragged-robins, once so pink,
> Now are turn'd as black as ink,

And the leaves, being scorch'd so much,
Even crumble at the touch.

The bright pink flowers of *Lychnis flos-cuculi*, called ragged robin, although its name translates literally as "the flower of the cuckoo," appear in Tennyson's poem "Marriage of Geraint" from *The Idylls of the King*. The unrefined flower represents a Cinderella, whisked from her lower-class background into the circles of nobility by the young and impetuous love of a prince:

Let never maiden think, however fair,
She is not fairer in new clothes than old.
And should some great court-lady say, the Prince
Hath pick'd a ragged-robin from the hedge,
And like a madman brought her to the court,
Then were ye shamed, and worse, might shame the Prince
To whom we are beholden.

Recalling the flame-like quality of the lychnis, Victorian poet Margaret Cecilia Furse in "The Lamp Flower" wants to follow the fairies on their lighted path:

The campion white
Above the grass
Her lamp doth light
Where fairies pass.

Softly they show
The secret way,
Unflickering glow
For elf and fay.

My little thought
Hath donned her shoe,

And all untaught
Gone dancing too.

Sadly I peer
Among the grass
And seem to hear
The fairies pass.

But where they go
I cannot see,
Too faintly glow
The lamps for me.

My thought is gone
With fay and elf,
We mope alone,
I and myself.

MARIGOLD

Spirits far more generous than ours

To the European gardener, the common marigold is *Calendula offici-nalis*, the pot marigold, a composite that traveled from its origin in the Mediterranean to England as early as the thirteenth century. In addition to being a commonly used bedding plant, *C. officinalis* has culinary and medicinal uses—evident by its species name, which means "of the pharmacy shops"—such as providing color to butter, thickening for soups, and healing for chilblains. The twenty annual and perennial species of calendula have daisy-like heads of yellow to yellow-orange, and the leaves are often aromatic.

Other marigolds include the cadmium-orange flowers of *Tagetes* from Mexico and Central America. This rather fetid-smelling genus consists of some fifty species, and hybrids of *T. erecta*, probably the most widely grown species in North America, have given rise to numerous cultivars. Another marigold is *Dimorphotheca*, seven species of sun marigolds native to Africa. The marsh marigold of England, *Caltha palustris*, might also be mentioned, but unlike the other members of the Compositae, it is in the Ranunculaceae. (The name *Caltha* is simply the classical Roman name for any plant with yellow flowers.)

The name *Calendula* is Latin for marigolds, because some species seem to flower almost continuously. The Romans noted how the plants remained in bloom through the months (even through mild winters), which were marked by the passing of the *calend* of each month. The *calendae* were the first days of each month on which interest on borrowed money was to be paid. Hence the marigold has been known as the flower of the calends.

The Aztec marigolds of the Americas, also called inappropriately the African or French marigolds, were named *Tagetes* by Linnaeus in honor of Tages, the grandson of Jupiter, Zeus to the Greeks, and an Etruscan deity. According to legend, Tages sprang from newly plowed earth and taught the Etruscans the omens regarding lightning, winds, and eclipses. The Spanish explorers brought the tagetes marigolds to Spain and northern Africa, where they flourished in Moorish gardens. The plants eventually arrived in England and were called marigolds because of their similarity to calendulas. In Mexico, the tagetes are referred to as *la flor de muerto*, the flower of death. They are put on graves on the Day of the Dead, or All Hallow's Day, 1 November, to venerate all departed ancestors and to remember the Indians killed by Cortés; the flower is said to have sprung from their blood.

The marigold is referred to in early English writings sometimes simply as "gold," with variant spellings such as "gowle" and "goule." Monks and nuns passed along the legend that the Virgin Mary wore a marigold on her bosom as a material symbol of the golden glow of light radiating around her head. By medieval times the flower was called Mary's gold, which later became marygold, and now marigold. Some English writers have also called the marigold Mary's-bud or Mary-buds. It is not clear exactly when marigolds began to be used for ecclesiastical purposes, but they have been employed to decorate churches for Ladytide on 25 March, the feast day in honor of the annunciation of the Virgin. One bit of lore even claims that the flowers are named after the Virgin Mary because they are in bloom for any and all feasts held in her honor. The flower's association with the Virgin Mary is reinforced in "The Village Patriarch" by Ebenezer Elliott, in which a blind mason named Enoch Wray reflects on the hard life he has lived in the rural countryside. The marigold "is the flower which (pious rustics say)/The Virgin Mother in her bosom wore."

In the language of flowers, marigolds suggested sorrow and despair and sometimes grief and misery. Communication using the language of flowers could have complicated shades of meaning; for example, in a bouquet of marigolds, the grief could be tempered by the addition of other flowers. Adding poppies to a tussie-mussie of marigolds changed their meaning to "I will soothe your grief." Mixed with other flowers, it generally marked the joys and sorrows of life. But on the other hand, to dream of marigolds portended prosperity and riches. No doubt the flower suggested the gold to be received by the dreamer. But Robert Herrick suggested in "How Marigolds Came Yellow" that their color shows jealousy:

> Jealous girls these sometimes were,
> While they lived, or lasted here;
> Turned to flowers, still they be
> Yellow, marked for jealousy.

John Gay, famous for his virulent political satire *The Beggar's Opera*, authored a series of six pastorals after Virgil called *The Shepherd's Week* (1714). His shepherds and milkmaids pose riddles to one another, such as they do in this section titled "Monday, or the Squabble": "What flower is that which bear's the Virgin's name,/The richest metal joined with the same?" The answer is clearly "Mary gold."

The poet and pamphlet writer George Wither of England wrote an observation of the marigold during his pastoral period in the early seventeenth century. He described the marigold flower constantly facing the sun, personified by Apollo, or Phoebus, through the course of the day:

> When with a serious musing I behold
> The grateful and obsequious Marigold
> How duly every morning she displays
> Her open breast when Phoebus spreads his rays;

How she observes him in his daily walk,
Still bending towards him her small slender stalk;
How, when he down declines, she droops and mourns,
Bedewed, as t'were, with tears till he returns;
And how she veils her flowers when he is gone,
As if she scorn'd to be looked upon
By an inferior eye, or did contemn
To wait upon a meaner light than him.
This I meditate, methinks the flowers
Have spirits far more generous than ours.

Marguerite of Orleans, grandmother of the French Bourbon dynasty's great King Henri IV, selected marigolds for her coat of arms, displaying the flowers with petals facing the sun. Her motto stated, "*Je ne veux suivre que lui seul*" (Follow him I will only). George Wither alluded to this heraldic use in "The Choice":

Let who list, for me, advance
The admired flowers of France,
Let who will praise and behold
The reserved Marigold.

The sunrise opening and sundown shutting of the marigold (probably often *Calendula*) were significant for many authors, including Shakespeare. In *Cymbeline* (1623), "winking Mary-buds begin/To open their golden-eyes." The marigold symbolizes a somewhat downcast mood in *The Winter's Tale* (1610): "The marigold that goes to bed with the sun,/And with him rises weeping." Such is the closing of the calendula and its dew-drenched opening after the sun rises. And William Browne of Tavistock recalled in "Memory," "So shuts the marigold her leaves/At the departure of the sun."

In the chancel of Devonshire's Berry Narbor Church is an epitaph to seven-year-old Mary Westwood who left "this vale of miserie for

a mansion in Felicitie . . . Januar. 31. Anno Domini. 1648." Young Mary
was compared to a marigold:

> This Mary-gold lo! here doth shew
> Marie worth gold lies near below;
> Cut doune by death, ye fair'st-gilt flour
> Flourish and fade doth in an hour.
> The Mary-gold in sunshine spread
> When cloudie clos'd doth bow the head,
> This orient plant retains its guise
> With splendent Sol to set and rise—
> Ev'n so this Virgin Marie Rose,
> In life soon nipt, in death fresh grows.

MARSH MALLOW

Of marsh-mallows my boat is made

The marsh mallow or white mallow, a shrubby herbaceous plant that grows naturally along wet, lowland areas, is *Althaea officinalis*, a member of the Malvaceae. *Althaea* is a genus of twelve species native from western Europe to central Asia; *A. officinalis* has become naturalized in the eastern United States. The althaeas typically have large palmate leaves and rose-pink blossoms that account for their cultivation as ornamental plants. *Althaea* is closely akin to *Alcea* (see the chapter "Hollyhock"), *Hibiscus*, and *Malva*, genera that at times have "shared" one an-

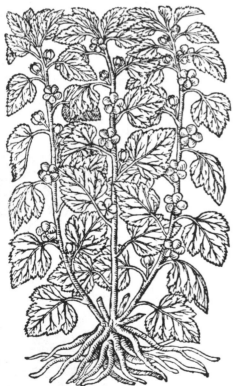

other's species through taxonomic and nomenclatural changes. In some regions the common name mallow applies equally to all four genera, but the species of *Malva* are generally considered the true mallows.

The name *Althaea* is from the Greek *althaine*, which means to heal or to cure, in reference to the medicinal properties of the most well-known species, *A. officinalis*. Drawing on the plant's innate healing ability, the marsh mallow represented maternal tenderness and beneficence in the language of flowers. The word *officinalis* means "from the apothecary" (the pharmacy office). The common name marsh indicates that this is the mallow that grows in wet areas. The word mallow appears to be derived from the Anglo-Saxon word *malwe*, which came from the Greek *malake*, for soft, a reference to the plant's mucilaginous root. The same Greek word led to the name of the mineral malachite, because of the fine green color it shares with the leaves of the mallow plant. *Malake* is also the etymological source of the genus name *Malva* and family name Malvaceae. Mallow was called "mersc mealwe" in the year 1000, a fact documented by T. Oswald Cockayne in 1864 in *Leechdoms, Wortcunning, and Starcraft of Early England.*

The medical properties of the marsh mallow have been extolled by Horace, Virgil, Dioscorides, and Pliny. The herb was also known and used by the Chinese and the Egyptians. The gummy roots were used to make a coating for the stomach and intestines, presumably for ulcers, to cure sprains and muscle aches, and to treat chest and bronchial problems. The plants were also served as a vegetable. In the Bible the book of Job (30:3–4) mentions a famine in which mallow was eaten: "For want and famine they were solitary; fleeing into the wilderness in former time desolate and waste. Who cut up mallows by the bushes, and juniper roots for their meat." Pythagoras, a Greek philosopher and mathematician of the sixth century BCE, advised abstaining from eating mallows because Mallow was the first messenger sent to earth by the gods as a token of their sympathy with mortals. Eating the mallow plant would dishonor the gods. More recently, the marsh mallow has been used as a confectionery paste and for making lozenges; however, contemporary store-bought marsh mallows are made of corn syrup, dextrose, and gelatin and contain no part of the marsh mallow plant.

The flat, round seed pod of the marsh mallow and its relatives has somewhat of a resemblance to a cheese. *Malva sylvestris*, the tall or high mallow that grows up to 1 meter (3 feet) high, is often simply called cheeses. Among the folk in some English counties, mallow was called fairy cheeses. John Clare recalled this folk name in a poem about eating the unripe seed:

The sitting down, when school was o'er
Upon the threshold of the door
Picking from Mallows, sport to please
The crumpled seed we call'd a cheese.

"Lament for Bion," an anonymously written Greek poem translated by E. J. Myers, compares the perennial mallow to the single-season life of a human:

Ay me, ay me, the mallow in the mead,
The parsley green, the anise-tendril's ring,
Fade all and die, but in due season freed
Grow yet again and greet another spring.

But we, we men, the mighty and the strong,
Wise-witted men, when our one life is o'er,
Low laid in earth sleep silently and long
A sleep that wins no waking evermore.

Chinese emperor and poet Ch'ien Wen-ti, reigning in the year 500, wrote "The Ferry," translated by Arthur Waley, about a boat made of the marsh mallow:

Of marsh-mallows my boat is made,
The ropes are lily-roots.
The pole-star is athwart the sky:
The moon sinks low.

Shakespeare mentioned mallows in *The Tempest* (1611), when Gonzalo, an honest old counselor, tells Antonio and Sebastian, "Had I plantation of this isle, my lord—" and he is interrupted by their saying, "He 'ld sow 't with nettle-seed. Or docks or mallow."

Mallow appears in *Thierry King of France* (1621), a play by Sir Francis Beaumont and John Fletcher. Legions of doctors attended the poisoned Thierry, including physicians from Spain, Italy, and

after him,
An English doctor with a bunch of pot-herbs,
And he cries out, "Endive and succory,
With a few mallow-roots and butter-milk!"

George Crabbe, English poet from Suffolk, wrote a long epistolary poem called *The Borough* (1810) in which he illustrated the life of a country town and its polluted pathways. He said of the poor district in letter 18, "Here the strong mallow strikes her slimy root/Here the nightshade hangs her deadly fruit." Earlier in 1810 in *The Village*, he described the rank weeds that grow in poor neighborhoods near rocky coasts. He observed, "Hardy and high, above the slender sheaf,/The slimy mallow waves her silky leaf."

Alfred, Lord Tennyson, in "The Brook" saw "Many a fairy foreland set/With willow-weed and mallow."

MORNING GLORY & BINDWEED

When the dawn brightens into joy

The morning glory is a member of the Convolvulaceae, a vast family of climbing and sprawling herbaceous and shrubby plants. The well-known morning glory of gardens is of the genus *Ipomoea*; members of the genus *Convolvulus* are also called morning glories or Indian bindweed. Of the approximately 250 species of *Convolvulus*, *C. tricolor* offers the bright 'Blue Ensign', the popular 'Heavenly Blue', and the carmine 'Crimson Monarch'. Of the approximately 500 species of *Ipomoea*, *I.*

nil is a climber to 5 meters (16 feet) and has selections that are brown ('Chocolate'), red ('Scarlett O'Hara'), and violet-purple ('Limbata'). The popular dark-leaved 'Blackie', chartreuse-leaved varieties, and variegated forms belong to *I. batatas*, the common sweet potato. *Ipomoea quamoclit* is the scarlet-flowered cypress vine. The ipomoeas are widely distributed in the tropics and subtropics, whereas the convolvuluses are primarily temperate to subtropical.

The name *Ipomoea* is from the Greek root words *ips*, meaning worm, and *homoios*, meaning like or resembling. The morning glory's viny, climbing habit is indeed suggestive of a worm. The name *Convolvulus* is from the Latin *convolva*, meaning twining or twisting.

Japanese mythology tells of the origin of morning glories, or in Japanese *asagao*, literally morning face. The sun goddess, who sat at her loom weaving colorful garments for her children, one day became so angry with her devilish brother that she shut herself up in a cave to get away from him. With the sun goddess in a cave, her bright light diminished and the Land of Many Islands, Japan, was bathed in perpetual twilight. The gods and mortals begged her to come out of the cave and called upon the god of thought for a suggestion. He commanded the 800 gods to collect 500 colored jewels and to hang them on a tree outside the cave. He then placed a huge mirror against the tree and entreated the gods to make great merriment amongst themselves. The joyous sounds reached the cave and made the sun goddess curious. Soon, she came out and was dazzled by the brightness of her light on the mirror and the jewels. The gods rejoiced, but in their excitement the mirror crashed down upon them, bringing the tree with it, burying the gods under broken glass and jewels. The next morning multicolored morning glories sprang up from the ground on the Land of Many Islands. Since then, morning glories have been called the Jewels of Heaven.

In English literature the names morning glory and bindweed are often interchanged and used without clear distinction of botanical classification. For example, species of *Calystegia* in the Convolvulaceae are also called bindweed. The same indistinctness was present in the language of flowers: the pink convolvulus stood for worth sustained by judicious affection, the great bindweed suggested insinuation, and the morning glory expressed affectation.

Bindweed and woodbine have been the subject of debate among Shakespearean scholars, arising possibly from a simple typo. It is generally accepted that woodbine is a synonym for honeysuckle. For example, Oberon says in *A Midsummer Night's Dream* that he knows a bank "Quite over-canopied with luscious woodbine," but in the same play Titania tells Bottom,

> Sleep thou, and I will wind thee in my arms
> Fairies, be gone, and be all ways away.
> So doth the woodbine the sweet honeysuckle
> Gentle entwist; the female ivy so
> Enrings the barky fingers of the elm.

Scholars have argued that Shakespeare meant to write "bindwood" in the above quote, referring to bindweed, since he did not likely mean to write that the woodbine (a k a honeysuckle) was entwisting upon honeysuckle. It is suggested that a printer transposed the syllables, assuming Shakespeare meant to write "woodbine." Another viewpoint is that "woodbine" means the vine only, and "honeysuckle" refers solely to the flowers.

Ben Jonson also wrote of both honeysuckles and morning glories. In one of his masques, "The Vision of Delight," a short play presented at court in 1617, the muses sing and dance. The character Delight sings of the joys of spring. Wonder speaks of the wealth of Nature and its flowers, represented on stage by the Bower of Zephyr. Wonder says at the sight of the bower:

> Behold
> How the blue bindweed doth itself infold
> With honeysuckle, and both these entwine
> Themselves with bryony and jessamine,
> To cast a kind and odoriferous shade.

John Parkinson described several "bindeweedes" and convolvuluses without recognizing that they were different plants. He noted of the "great blew bindeweede": "This goodly plant riseth up with many long and winding branches, whereby it climbeth and windeth upon any poles, herbes, or trees, that stand neare it within a great compasse, alwaies winding itselfe contrary to the course of the sunne."

In *The Herbal or General History of Plants* (1633 revision by Thomas Johnson) John Gerard described several bindweeds and morning glories and showed rich illustrations of vines, fleshy roots, and flowers. He advised that in order to grow them, "The seed must be sowne as melons and cucumbers are, and at the same time; it floured with me at the end of August." He went on to quote a couplet by the Roman Columella, translated from Latin: "In baskets bring thou violets, and blew bindweed withall,/But mixed with plesant baulme and cassia medicinall."

In "The Scholar-Gypsy," a poem about an Oxford student who leaves his studies for financial reasons, going to live with and learn from gypsies, Matthew Arnold noted the growth habit of the bindweed:

> Through the thick corn the scarlet poppies peep,
> And round green roots and yellowing stalks I see
> Pale pink convolvulus in tendrils creep;
> And air-swept lindens yield
> Their scent . . .

A poem quoted in *Century Magazine* of December 1897 describes the joy of seeing a morning glory:

> Every morn, when the dawn brightens into joy
> The morning-glory renews its beautiful flowers,
> And continues blooming long in this way,
> To give us hope and peace that wither not.

Finally, Walt Whitman succinctly described the effect of the flower in "Song of Myself" (1855): "A morning glory at my window satisfies me more than the metaphysics of books."

NASTURTIUM

A saint-like glory trembles round her head

The garden nasturtium or Indian cress is a member of the genus *Tropaeolum* in the Tropaeolaceae. The distribution of its eighty-five species is primarily in Central and South America, from Mexico to Brazil, the Argentine Patagonia, and Chile. *Tropaeolum majus* is a climbing annual. Among its numerous cultivars are 'Variegatum' (or 'Alaska'), with flame-orange flowers, and 'Peach Melba,' with creamy yellow flowers and a blotched throat. *Tropaeolum peregrinum*, the canary creeper, has sulfur-yellow flowers, reminiscent of a flock of canaries; when the Spanish saw it in Peru they called it *la flor de pajarito*, the little bird flower. *Tropaeolum tuberosum*, also from Peru, has edible potato-like tubers and showy orange flowers.

Linnaeus took the genus name *Tropaeolum* from the Greek *tropaion*, meaning trophy. The Latin equivalent is *tropaeum*. Its climbing stems and large leaves, Linnaeus thought, looked like a pillar used for displaying battlefield armor from conquered foes. He also thought the flower resembled a spear-pierced golden helmet with blood stains. The common name nasturtium is from *nasus torquere*, meaning "to twitch the nose," the reaction to a whiff of the pungent peppery aroma of the plant. The appellation Indian cress was provided by Europeans because it came from the New World and because the spicy taste of the leaves and flowers was reminiscent of watercress, *Nasturtium officinale*, a member of the Cruciferae. John Evelyn, in *Acetaria, a Discourse of Sallets* (1699), remarked that "above all the Indian [cress], moderately hot, and aromatic, quicken the torpent Spirits, and purge the Brain, and are of singular effect against the Scorbute [scurvy]." He added that "the buds being candy'd, are likewise us'd in Strewings all Winter."

In French the nasturtium is *capucine* and in Italian *cappucina*—both are related in a round-about way to the beverage cappuccino. The French and Italian names come from *cappuccio*, meaning hood or cowl, which the nasturtium flower was thought to resemble. An order of Franciscan monks, the Capuchins, wore drab habits with a simple pointed hood of an earthy chocolate color. The Capuchins believed that these simple cloaks most closely resembled the ones that St. Francis originally wore. The drab color of the habits, similar in color to coffee, was the basis for the name cappuccino. In English, a variation of the name, capucine capers, was used for the somewhat edible pickled fruit of nasturtium.

In its native land of Peru, the nasturtium was called gold nugget by the Incas because of an Andean legend. In the days of the Spanish conquistadors, a converted Incan given the Christian name Juan decided to replace the gold that the Spaniards had looted from the village temple. Juan went to a hidden spring to look for gold that may have been washed up by recent floods. His search was rewarded with a hefty bag of gold. As he returned with his bounty, some Spanish riders spotted him and attacked. Juan prayed to the mountain god to save him as he tossed the bag of gold into the woods. A Spaniard scrambled to recover the bag and was bitten by a poisonous snake sent by the mountain god; instead of gold he found a spread of golden-colored flowers of nasturtium. Later, Juan recovered the gold nuggets, reinstated by their mountain god, and took them back to his village. There the local smith beat the gold into thin sheets to make the temple's idols.

Nasturtiums first reached English gardens in the 1570s as *Tropaeolum minus*, a dwarf annual. Debate continues over when *T. majus* arrived in Europe from South America; some suggest the 1680s, but John Parkinson in his *Paradisi in Sole, Paradisus Terrestris* of 1629 described a plant suggestive of *T. majus*, which he called yellow lark's heels. Parkinson was effusive: "This flower . . . is of so great beauty and sweetnesse withall, that my Garden of delight cannot bee unfurnished of it." He also recommended using it in a tussie-mussie or nosegay by placing the nasturtium in the middle of carnations or gilliflowers.

Linnaeus's daughter, Elizabeth Cristina, reported to her father that during the summer she could see sparks or flashes of light at sunrise and sunset, after and before total darkness, respectively, emanating from the nasturtium plants. Her observation became the subject of much scientific and philosophical debate, and Erasmus Darwin alluded to it in his long poem *The Loves of the Plants* (1789):

> The chaste Tropaeo leaves her secret bed,
> A saint-like glory trembles round her head;
> Eight watchful swains, along the lawns of night,
> With amorous steps pursue the virgin light;
> O'er her fair form the electric lustre plays,
> And cold she moves amid the lambent blaze.

A celebrated scientist of the time declared these scintillations to be electric. But in a footnote to the above passage, Darwin noted, "This curious subject deserves further investigation."

The Jesuit poet René Rapin saw heroic young soldiers in the nasturtiums:

> Shield-like Nasturtium, too, confusedly spread,
> With intermingling Trefoil fill each bed—
> Once graceful youths; this last a Grecian swain,
> The first an huntsman on the Trojan plain.

ORCHID

That flowret's velvet breast

An astounding 800 genera of orchids consisting of nearly 18,000 species make up the Orchidaceae. Distributed worldwide, the species may be epiphytic, meaning air-growing, or terrestrial; some are hardy and others are frost-tender. One family member, *Vanilla planifolia*, is the source of vanilla flavoring.

The genus *Orchis* is named from the Greek word for testicles, alluding to the shape of the paired round tubers on some species; they are often of different sizes, with one storing the food from the previous year's growth. Hence, according to herbalists using the Doctrine of Signatures, orchids are testicle analogs and thus have aphrodisiac properties. Orchid tubers were sometimes called dogstones, hare's ballockes, and sweet cods. Likewise, the name vanilla was applied because of the long bean pods of the vanilla orchid; in Spanish the pods and plants are called *vaina*, a word derived from the Latin word *vagina*, meaning sheath.

Considered by many to be the elite flower of the world, the orchid has taken many common names. The early purple orchid, *Orchis mascula*, a terrestrial plant of Europe, was called Gethsemane because its purple-spotted leaves suggested the Passion of Jesus; according to legend it grew on Calvary, the place of his crucifixion:

> Those deep unwrought marks
> The villager will tell you
> Are the flower's potion from the atoning blood
> On Calvary shed. Beneath the cross it grew.

This and other folktales from England indicate that the purple orchid's spots suggest blood. Such legends were the source of the common name bloody-butcher. The spotted orchid, formerly *Orchis maculata* but now *Dactylorhiza maculata* or *D. fuchsii*, is believed to be the "long purple" mentioned by Shakespeare in *Hamlet* (1601). Also in *Hamlet*, when Queen Gertrude tells Laertes that his sister has drowned, she describes the dead Ophelia's garland of flowers, including two names for the spotted orchid:

> There with fantastic garlands did she make
> Of crowflowers, nettles, daisies, and long purples
> That liberal shepherds give a grosser name,
> But our cold maids do dead men's fingers call them.

The orchid appears in Greek mythology as well. Orchis was the son of the nymph Acolasia and the satyr Patellanus, whose union resulted in unbridled passion and excess. Their wanton son, Orchis, presided over feasts honoring Priapus, who represented the male procreative power, and Bacchus, the god of wine and revelry. At one celebration, Orchis allowed his hands to wander freely onto one of the priestesses. Such an action was sacrilegious, and the gods were so enraged that they pulled his limbs from his body and tore him into small pieces. The grieving Patellanus transformed the remains of his son into the orchid flower. Another version of the myth indicates that semen from Patellanus's copulations produced the satyrion orchid, "the flower of the satyrs," now recalled in the genus *Satyrium* that grows in subtropical areas. This association with virility is the basis for the term satyriasis, a male's abnormal and uncontrollable desire for sexual intercourse. Pliny the Elder believed that merely holding an orchid tuber would ignite sexual desire, particularly if the tuber were dipped in a potion of wine.

The North American and Eurasian woodland orchid, *Cypripedium*, is called moccasin flower or lady's slippers because of its resemblance to a small shoe. The name *Cypripedium* was selected by Linnaeus in

1737; it derives from Cyprus, the mythological island birthplace of Aphrodite, and from *pedilum*, for shoe or foot.

In the four cantos of *The Loves of the Plants* (1789) Erasmus Darwin explained the Linnaean system of plant classification through the voice of the goddess of botany. She descends to earth to teach that the members of the "vegetable world" are classified in part according to the number of stamens, pistils, and other floral parts. The goddess makes an analogy between the flower parts and (floral) harems of beaux and belles and marriage beds. For example, *Colchicum autumnale* with three styles (pistils) and six stamens is described as "Three blushing maids the intrepid Nymph attend,/And six gay youths, enamoured train! defend." About the orchid, she says,

> With blushes bright as morn fair Orchis charms,
> And lulls her infant in her fondling arms;
> Soft plays affection around her bosom's throne,
> And guards his life, forgetful of her own.
> So wings the wounded deer her headlong flight,
> Pierced by some ambush'd archer of the night,
> Shoots to the woodlands with her bounding fawn,
> And drops of blood bedew the conscious lawn;
> There, hid in shades, she shuns the cheerful day,
> Hangs o'er her young, and weeps her life away.

To this section on the orchid Darwin added a footnote explaining *Orchis* tuber formation: "The orchis morio, in the circumstance of the parent-root shriveling up and dyeing, as the young one increases."

The language of flowers declared that orchids in general suggested beauty. The bee orchid, *Ophrys apifera*, has flowers that look like bees glued to a stalk, which would be an error of nature; hence, in the language of flowers the bee orchid indicated error. English poet John Langhorne of the mid-eighteenth century described this unusual plant in "The Bee Flower" from his *Fables of Flora*:

See on that flowret's velvet breast,
How close the busy vagrant lies!
His thin-wrought plume, his downy breast,
The ambrosial gold that swells his thighs.

Perhaps his fragrant load may bind
His limbs; we'll set the captive free;
I sought the living bee to find,
And found the picture of a bee.

The mimicry across plant and animal kingdoms that Langhorne alluded to is one of nature's tricks, repeated in many other orchids and plants in general. John Gerard noted the phenomenon in his descriptions of the different kinds of orchids, which he called fox stones:

Some have flours wherein is to be seen the shape of sundry sorts of living creatures, some the shape and proportion of flies, in other gnats, some humble bees, others like unto honey bees; some like butter flies, and others like waspes that be dead; some yellow of colour, others white; some purple; mixed with red, others of a browne over-worn colour.

PEONY

Whose blushes might the praise of virtue claim

The thirty-three species of *Paeonia*, a shrubby or herbaceous perennial, are native to the Northern Hemisphere in the temperate regions
of Europe, Asia, and western North America. The ornamental forms
are characterized by large showy flowers that vary from white to yellow and red. Some species have been grown in Japan and China for
hundreds of years. *Paeonia lactiflora* is the species from which many of
the hybrid peonies of gardens were derived. The Moutan peonies (*P.
suffruticosa*), some of which are double forms, originated on the Mou-
Tan-Shan mountain slopes in China and Tibet and are the source of
the shrubby "tree" peonies that may grow up to
2 meters (7 feet) tall. The word Moutan means
"most beautiful," referring to the peonies
for which the mountain is named. *Paeonia californica* and *P. brownii* are two species
of western North America.

The name *Paeonia* has the familiar etymological root *paean*, an epithet or hymn
of praise to a helping god. But according
to mythology, the Latin name *Paeonia* is
derived from Paeon, a physician of the
gods in Greek mythology, who reputed-
ly was the first to use the plant's wondrous
medicinal attributes. His name generally
has been used to mean healer. Paeon, a
son of Endymion, lost a race at Olympia.
Disappointed by his defeat, he went into exile in Mace-
donia where he founded a race of people bearing his

name, the Paeonians. Paeon, according to Homer, healed Hades after Heracles wounded him in the Trojan War. Another Greek healer, Asclepius, was jealous because Paeon had been asked to cure Hades. When Asclepius devised a plot to kill Paeon, Hades learned of it and intervened. In gratitude to Paeon for healing him, Hades protected the healer by turning him into a flower that now bears his name.

A different version of the origin of the plant's name says that Asclepius, the son of Apollo, was reared and trained by the centaur Chiron. By the time Asclepius had grown to manhood, he had become the physician of the gods. Because of his great knowledge of herbs and healing, he was called Paeon, meaning helper. Thus, the early doctors were called paeoni because of their affinity with that early healer and because of the use of the peony plant in their practice. It is an important ingredient in Chinese medicine.

An additional legend relating to the origin of the plant suggests that Paeonia was a beautiful nymph with whom Apollo often flirted. One day Aphrodite caught them together. The shy Paeonia blushed so red that the color never left her face when the angry Aphrodite changed her into a rosy-colored peony.

In *De Hortorum Cultura* (1665), translated by James Gardiner (1706), René Rapin used the peony to portray the pride of Apollo's lust:

> Erect in all her scarlet Pomp you'll see
> With busy leaves the graceful Peony;
> Whose Blushes might the Praise of Virtue claim,
> But her vile Scent betrays they rise from Shame.

> If while Alcinous bleating Flock she fed,
> A heav'nly Lover had not sought her Bed,
> Shy of Mankind her Pride preserv'd her just,
> And Pride betrayed her to Apollo's Lust.

Peonies were highly regarded in China, where *Paeonia suffruticosa*, the Moutan or tree peony, was the queen of flowers, signifying love and reverence. One Chinese legend tells of a young scholar who grew numerous flowers, including the peony. One day he was visited by a young maiden who admired his garden. She was hired as a servant and soon became a companion and lover. On the day of a scheduled visit from a moralist, the young scholar could not find his love. He searched and searched and finally found in a gallery in his home a shadowy specter of his lover fading into the wall. She told him that she was the soul of the peony, who had been warmed into human shape by his friendship and love. But, she said, the moralist would not approve of their relationship and she must return to the flowers. The scholar went into deep mourning as he continued to tend his garden and search for her return, but he never saw her again.

The language of flowers echoed the association between the peony and an uncondoned relationship—the peony represented shame and bashfulness to the Victorians. But to the Japanese, where it was called the flower of June, it suggested happiness, marriage, and prosperity. The Chinese, too, saw the peony as the flower of prosperity.

In literature, peonies made an appearance as early as the eighth century in the poems of Chinese poet Po Chü-I, a governor who held minor posts and whose poetry provoked the criticisms of officials for its depictions of the sufferings of town folk. In "The Flower Market," translated by Arthur Waley, he wrote,

> In the Royal City spring is almost over:
> Tinkle, tinkle—the coaches and horsemen pass.
> We tell each other 'This is the peony season.'
> And follow with the crowd that goes to the Flower Market.

Another Chinese poet, Li Po, was ordered by the emperor to compose poetry for the emperor's wife. He wrote, "She is the flowering branch of the peony,/Richly laden with honey dew" (translated by Shigeyoshi Obata).

Several European superstitions are associated with the peony. Because it originated from the moon goddess, Selene, the plant was to be collected only at night, lest Picus, the woodpecker guarding it, attack the eyes of those attempting to collect the plant during the day. Because it was so risky to collect the roots of the plants, the safest way was to tie a hungry dog to a peony and get him to pull out the plant by luring him away with roasted meat. If left to grow, the peony was said to shine at night, guarding shepherds and their sheep.

The association of peonies with the moon suggested it as a remedy for lunacy (an old name for the peony was *Rosa Lunaria*), as well as nervousness, epilepsy, and liver obstructions, among other things. Josuah Sylvester, in *Du Bartas' Triumph of Faith, the Sacrifice of Issac* (1591), provided an additional medicinal use: "About an Infants neck hang Peonie, It cures Alcydes cruell Maladie." Even in the nineteenth century, the Sussex folk made strings of peony root beads for teething children and amulets for warding off illness and evil spirits.

Regarding peonies John Gerard said in *The Herbal or General History of Plants* (1633 revision by Thomas Johnson) that "The black graines (that is the seed) to the number of fifteene taken in wine or mead, helpes the strangling and paines of the matrix or mother, and is a speciall remedie for those that are troubled in the night with the disease called Ephialtes or night Mare." The seeds of *Paeonia officinalis* and *P. corallina* were sometimes used as a spice. In *Piers Plowman*, the Middle English poem written in the 1360s and attributed to William Langland, Gloton (or Glutton) is on the way to "holi church for to here masse" one Friday morning when Betene (or Betsy), the keeper of the tavern, invites him in for a drink. Gloton asks if she has any "hote spices" to season a drink and Betene replies, "I haue peper and piane [peony] and a pound of garlek." This is an offer that Gloton cannot refuse; he delays his trip to the church where he would "synne na more" and passes several hours with Betene.

The distinction between male and female peonies, sometimes referred to in old writings and originating with Dioscorides in the first century, has no obvious connection with the sexes of the flowers. The two "varieties," as they were known, are based on the roots of different species: the female *Paeonia officinalis* has roundish, thick knobs or tubers, and the male *P. mascula* (formerly *P. corallina*) has several oblong knobs. William Turner, writing in *A New Herball* (1562), observed: "The female [peony] is common throughout all England and Germany; and in divers places in England and in some parts of Brabant (as in Peter Coddenberg's garden in Antwerp) the male groweth also; but I could never see it in high Germany. The fairest that I did see was in a rich clothier's garden."

In Shakespeare's masque in act 4 of *The Tempest* (1611), Iris greets Ceres, or Demeter, the goddess of grain and harvest: "Thy banks with pioned [peonied] and twilled [lilied] brims/Which spongy April at thy hest betrims/To make cold nymphs chaste crowns."

"Ode to Melancholy" shows that John Keats knew peonies could provide treatment for melancholy:

> But when the melancholy fit shall fall
> Sudden from heaven, like a weeping cloud
> That fosters the droop-headed flowers all,
> And hides the green hill in an April shroud;
> Then glut thy sorrow on a morning rose
> Or on the wealth of globed peonies.

The wild peony was singularly praised by the Rev. William Lisle Bowles, vicar of Bremhill and canon at Salisbury in the first half of the nineteenth century:

> One native flower is seen—the peony—
> One flower, which smiles in sunshine and in storm,
> There still companionless, but yet not sad,
> She had no sister of the summer field,
> None to rejoice with her when spring returns,

None that in sympathy may bend its head
When evening winds blow hollow o'er the rock in
 Autumn's gloom.

Despite its rich history in legend and considerable use as an herbal in both the East and West, the peony is today considered an old-fashioned, grandmother's flower—the "piney," as it was called in the southern United States. It was among the first flowers that European colonists brought to North America. Perhaps it is time to compose a paean to the lovely but shy nymph Paeonia and her namesake flower, which an anonymous poet depicted "With gilly flowers all set round,/ And pyonys powdered ay betweene."

PERIWINKLE

Flinging their light of blossoms

The most well known and widely cultivated of the six species of the periwinkle genus *Vinca* are *V. minor* and *V. major* in the Apocynaceae or dogbane family. Distributed in Europe, northern Africa, and Central Asia, the species are low-growing herbs or subshrubs, often evergreen and trailing. The handsome blue color of the common vinca of Europe, *V. minor*, occasioned John Parkinson to describe it as "a pale or bleake blew colour."

The genus name derives from the Latin *vincio*, meaning to bind, presumably referring to its vining habit or the use of the foliage in wreaths and decorations. The modern English common name has its origins in the Middle English *per wynke*, which comes by way of the Latin *vinca pervinca*. An obsolete meaning of the word periwinkle used in the fourteenth century is one who surpasses or excels, the fairest, or the "pink of perfection." *The Romance of Sir Degrevant* is a romance of the late fourteenth century from the North Midlands of England. Of unknown authorship, containing rich descriptions of feudal society, the poem tells the story of Degrevant's wooing of Melidor. He says to her,

"Corteys lady and wyse,/As thou arte pervenke of pryse/ I do me on thi gentryse,/Why wolt thou me spyll?" Similarly in Old French *pervenke* meant the finest or choicest, as in "*De tous vins ce est le pervenke*" (It is the periwinkle of all wines).

Nicholas Culpeper wrote that the plant belongs to Venus and that the leaves, if eaten together by a man and a woman, cause love to blossom between them. Thus it was considered an aphrodisiac that should be grown in the gardens of newlyweds. In the language of flowers, the blue periwinkle suggested pleasures of memory and the white noted pleasant recollections and sincere friendship. But the periwinkle was also associated with death—a former Italian name was *fiore di morte*, or flower of death. In medieval England periwinkle was made into garlands to place on the heads of condemned persons on their way to the gallows. The blue flowers were also used in the funeral wreaths and on the biers of deceased children.

The plant is called perevinke in *Romaunt of the Rose*, in which several allegorized figures fall in love with a rose. The *Romaunt of the Rose*, though much of it was translated by Chaucer, is based on the medieval French poem *Roman de la Rose* by Guillaume de Lorris. The version of the tale most often used is a translation from the 1500s that combines the works of Chaucer and de Lorris. The narrator enters the Garden of Mirth in a dream and describes the various figures he meets. In the following excerpt, he first comes upon the clothes of Cupid, then a fountain, "welles," beside which violets and periwinkles grow:

> His garments were every dele
> Iportraied and wrought with floures,
> By divers medeling of coloures;
> Floures there was of many a gise,
> Iset by campace in a sise;
> There lacked no floure to my dome,

> Ne not so moch as floure of brome,
> Ne violet, ne eke perevinke,
> Ne floure noen that men can think.
>
> . . . .
>
> Ther sprang the violete al newe
> And fresshe perevinke, riche of hewe,
> And floures yelowe, whyte, and rede;
> Swich plentee grew ther never in mede,
> Ful gay was al the ground, and quenyt,
> And poudred, as men had it peynt,
> With many a fresh and sondry flour,
> That casten up ful good savour.

The English romantic poet William Wordsworth wrote deeply personal thoughts of nature and "what man has done to man" in "Lines Written in Early Spring" (1798), in which the periwinkle has its place:

> I heard a thousand blended notes,
> While in a grove I sate reclined,
> In that sweet mood when pleasant thoughts
> Bring sad thoughts to the mind.
>
> . . . .
>
> Through primrose tufts, in that green bower,
> The periwinkle trailed its wreaths;
> And 't is my faith that every flower
> Enjoys the air it breathes.

Mary Russell Mitford in *Our Village* (1832), a series of sketches of English country life, made reference to periwinkle in her characteristically sunny and cheerful style:

> Ah! here is the hedge along which the periwinkle breathes
> and twines so profusely, with its evergreen leaves shining
> like the myrtle, and its starry blue flowers. It is seldom

found wild in this part of England; but when we do meet with it, it is so abundant and so welcome,—the very robin-redbreast of flowers, a winter friend.

The periwinkle charmed many a writer with its jaunty character. *The Cornhill Magazine* (November 1866) described "White periwinkles, flinging their light of blossoms and dark glossy leaves down the swift channels of the brawling streams." Also in the late nineteenth century, English poet laureate Alfred Austin, editor of the *National Review*, wrote of a location "Where along the hedgerow twinkle/Roguish eyes of periwinkle."

PHLOX

Smell sweet in the dusk

With the possible exception of one Siberian species that strayed from Alaska, the nearly seventy recognized species of *Phlox* are native to North America. These members of the Polemoniaceae may be annuals or perennials and herbaceous to shrubby. The flowers have five-parted, salver-shaped petals whose colors vary from white to purple. Species include the great autumn phlox of eastern North America, *P. paniculata* (formerly *P. decussata*), which grows to a meter (3 feet) tall and has given rise to numerous garden cultivars; *P. subulata*, the moss pink; *P. carolina*, the thick-leaf phlox; and *P. drummondii*, the annual phlox of Texas and the central United States. Many species are tall and straggly, taking up a lot of space in the garden border.

Linnaeus borrowed the name *Phlox* from the Greek word for flame, an allusion to the showy, vivid blush of color on the flowers of some species. Earlier, the phlox had been called *Lychnidea*, Greek for lamp, by Leonard Plukenet in an herbal of 1691. Because the North American–native phlox could not have been known at the time, references by Pliny and Theophrastus to a phlox-like plant were probably to a *Silene* (campion) or *Lychnis*. The word phlox is related to the Greek word for phlegm, which was believed by the ancients to be the cause of sore throats and inflammations.

Phlox made their way from North America to England in the early 1700s and were grown at the physic garden at Chelsea. Later, Victorians admired them and used phlox widely in bouquets; their pleasant fragrance suggested sweet dreams and love and became an implication of a marriage proposal. In the language of flowers they suggested unity and unanimity in reference to the close grouping of the flowers that

comprise the heads. Phlox were not generally grown as garden ornamentals in their native land until reintroduced from Europe.

John Bartram of Philadelphia sent his English correspondent Peter
Collinson, his "brother of the spade," "One sod of creeping Spring
Lychnis," that is, the rock garden species *Phlox subulata*, which prompted Reginald Farrer to write years later that the date of the gift, 10 December 1745, should be a horticultural holiday.

The plant that Bartram found and sent to England had already been
memorialized in Native American mythology. According to legend, a
Chickasaw tribe trespassed on the hunting grounds of the Creek Indians near the Savannah River in the southern United States. After a
battle of three days, the Chickasaw tried to retreat, but the Creek set
a fire wall through which the Chickasaw could not escape. Chuhla
(Blackbird), a little Chickasaw boy, summoned his animal friends to
help stamp out the fire. While fighting the fire, the white coat of the
squirrel became forever gray with ashes, the raccoon's tail became
ringed with pine resin, and the deer lost its tail. The Great Spirit, seeing the animals fight the fire, turned the embers to a blaze of fire-col-
ored flowers. The Creek arrived just in time to see smoldering stumps
and a wide path of green carpet containing the flaming flowers of the
moss-pink, *Phlox subulata*.

It has been said that phlox take up more space in the border than
they do in literature; nevertheless, William Cullen Bryant in "The
Maiden's Sorrow" (1856) wrote, "There, in the summer breezes, wave/
Crimson phlox and moccasin flower." In Edward Phillips's dictionary
titled *The New World of English Words* (1706), phlox is defined as "a Flower
of no Smell, but of a fine Flame-colour." In Mrs. Humphry Ward's
Story of Bessie Costrell (1895), "Phloxes and marigolds grew untidily about
their doorways and straggly roses, stained a little by the chalk soil,
looked in at the lattice windows." And Fredegond Shove described
garden phlox in "The Water-mill":

The phlox in the garden-beds
Turn red, turn grey
With the time of day,
And smell sweet in the dusk, then die away.

PIMPERNEL

In fiery red the sun doth rise

Linnaeus gave to the pimpernel the name *Anagallis*. Of the twenty or more species of these herbaceous members of the Primulaceae, only a few are grown as ornamentals. They are cosmopolitan in distribution, and some species, particularly *A. arvensis*, have become naturalized outside their native areas. *Anagallis arvensis*, the common or scarlet pimpernel, habitually opens promptly in the morning and closes by early afternoon, but since it closes when the weather threatens rain and may not open at all on cold and dark days, it is known as poor man's weather glass, shepherd's weather glass, and shepherd's clock or sundial. It is a plant to be noticed on bright sunny mornings when its red flowers shine at their best. Other garden species include *A. monellii*, the blue pimpernel, and *A. tenella*, the bog pimpernel, with sweetly scented pink flowers.

Anagallis is a Latinized version of the Greek *anagelao*, which means to laugh or to delight. It was a name given by the Greeks to plants that they believed cured sadness and melancholy through an infusion prepared from their leaves. Apparently Linnaeus selected the name to fit the plant's sunny disposition. The name pimpernel is related to the Old French *pimprenele*, the Spanish *pimpinela*, and the Latin *pipinella*—all corruptions of *bipinnella* or *bipennis*, meaning two-winged, a reference to its two tiers of leaflets in its compound leaf.

The following prayer implied that the scarlet pimpernel would prevent witchcraft. This prayer, recorded in an old manuscript in the Old Chetham Library, Manchester, would repel witches if uttered twice a day for a fortnight plus one day:

> Herbe Pimpernell, I have thee found,
> Growing upon Christ Jesus' ground:
> The same guilt the Lord Jesus gave unto thee,
> When He shed His blood on the tree.
> Arise up, Pimpernel, and goe with me,
> And God blesse me,
> And all that shall were thee. Amen.

In the language of flowers, the pimpernel connoted both change and the appointing of a rendezvous. Its medicinal properties involved the ability to instigate changes in the skin and eyes. John Parkinson wrote that "the distilled juice or water are by the French dames accounted merveillous to clear the skinne from roughness, deformity or discolouring thereof." Its leaves were believed to cure eye complaints and the bite from a mad dog. John Gerard thought the red form of the pimpernel was the female and the blue form the male and that the "juice purgeth the head by gargarising or washing the throat" and that "it helpeth those that be dim sighted—the juice mixed with honey cleanses the ulcers of the eye." In the Doctrine of Signatures, the red color of the pimpernel recommended its use to staunch blood. Such diverse virtues are noted in an old proverb: "No heart can think, no tongue can tell/The virtues of the pimpernel."

Erasmus Darwin in *The Loves of the Plants* (1789) included the pimpernel in his explanation of the Linnaean system of classifying the plant kingdom. Darwin noted the horological and meteorological attributes of the plant:

> Closed is the pink-eyed pimpernel;
> In fiery red the sun doth rise,
> Then wades through clouds to mount the skies;
> 'Twill surely rain—we see't with sorrow,
> No working in the fields to-morrow.

The prognosticating ability of the pimpernel is further lauded by an anonymous poet:

> And if I would the weather know, ere on some pleasure
>   trip I go,
> My scarlet weather-glass will show whether it will be fair
>   or no,
> The blue-eyed pimpernel will tell, by closed lids of rain
>   and showers;
> A fine bright day is known full well, when open wide it
>   spreads its flowers.
> Some flowers put on more gay attire, and this in
>   usefulness excel.
> But I, a shepherd, most admire the blue-eyed scarlet
>   pimpernel.

The Harvard professor of medicine Oliver Wendell Holmes wrote considerable light verse that was collected in numerous volumes. His work includes the poem "Pimpernel," despite the fact that there are no native pimpernels in North America:

> Some years ago, a dark-eyed maid
> Was sitting in the shade—
> There's something brings her to my mind
> In that young dreaming maid—
> And in her hand she held a flower,
> A flower whose speaking hue
> Said, in the language of the heart,
> "Believe the giver true."

And as she looked upon its leaves,
The maiden made a vow
To wear it when the bridal wreath
Was woven for her brow.
She watched the flower, as, day by day,
The leaflets curled and died;
But he who gave it never came
To claim her for his bride.

POPPY

Like a yawn of fire

In the Papaveraceae, the genus *Papaver* consists of fifty or more annual, biennial, or perennial herbaceous plants distributed worldwide except in South America. Many of the species are grown as ornamentals, and one species, *P. somniferum*, is the source of opium, a narcotic derived from the latex that oozes from the unripe seed pods when slit. The seeds are used as a source of cooking oil, in baking cakes and breads, and as birdseed. The Iceland or Arctic poppy, *P. nudicaule*, grows as tall as 50 cm (20 inches) and produces flowers of various colors. Widely known is the blood-red *P. rhoeas*, the corn or Flanders poppy, which has become a symbol of the soldiers who died during World War I. The oriental poppies, *P. orientale*, are bright orange, pink, white, and red. Petals of *P. orientale* and *P. rhoeas* are a source of dye used in wines and medicines. The popular California poppy is *Eschscholzia californica*, which is in the same family, though a different genus.

The common name poppy is derived from the Old English *popple*, itself having been derived from the Old French *pavau*, and ultimately the Latin *papau*. The name of the genus *Papaver* is Latin for milk, which comes from the Latin *pappa*, for breast milk, an allusion to the milk-like latex of some species' seed pods. In Spanish the name of the poppy is *adormidera*, for sleepy, a reference to the sleep-inducing properties of some species; the name is kin to the word dormitory, a place where one sleeps. In Italian, the name of the poppy is *alto papavero*, or literally "a tall poppy," an expression that suggests importance or high stature. Similarly, in English, a person of rank or distinction is sometimes called a tall poppy. John Milton compared the prestigious poppy to the common corn plant in "The Reason of Church-Government Urg'd Against Prelaty" (1641): "He little dreamt then that the weeding-hook of reformation would after two ages pluck up his glorious poppy from insulting over the good corne."

Greek mythology includes the poppy in the tragic story of Orpheus and Eurydice. Book 4 of Virgil's *Georgics* tells of Aristaeus, a shepherd or rustic deity, who had pursued Eurydice. He lost a swarm of bees through an epidemic. In order to have the hive restored, he had to sacrifice four bulls and four heifers to make peace with the shades—the souls that separate from the body after death. His mother, Cyrene, speaks thus to Aristaeus:

> When the ninth day has dawned
> You shall send oblivion's poppies as a funeral gift
>   to Orpheus,
> Slay a calf in honour of Eurydice placated,
> Slaughter a black ewe and to the thicket again.

Aristaeus obeyed his mother, and a living swarm of bees issued from the bellies of the rotting cattle.

Inspired by the poppy, Chaucer noted in *The Knight's Tale*, "A clarree, maed of a certeyn wyn/With narcotikes and opie of Thebes fyn." A clarree was a sweet liquor made of wine, clarified honey, and various spices. The alleged narcotic effect of the poppy was also mentioned by Shakespeare in act 3 of *Othello* (1622). Iago, intending to further provoke Othello concerning the seeming unfaithfulness of his wife, Desdemona, reminds Othello that he will never be able to go back to his innocent trust in her:

> Not poppy, nor mandragora,
> Nor all the drowsy syrups of the world,
> Shall ever medicine thee to that sweet sleep
> Which thou owedst yesterday.

And Frenchman René Rapin penned elegant praise to the poppy in appreciation of both the by-product and the beauty of the flower:

> And Poppy will erect her tufted Head,
> And earth be with a thousand beautys spread
> In this one Flow'r, which she to Ceres owes,
> Her self with much of wealthy Pride she shews;
> For some sublime a martial Scarlet stains,
> And some all silver'd o'er enrich the Plains:
> The pow'rful Seeds when press'd afford a Juice
> In Med'cine fam'd, and Sov'reign for its use;
> Whether in tedious Nights with charm to rest,
> Or bind the stubborn cough, and ease the lab'ring breast.

The English journalist and poet Leigh Hunt was a prolific writer imprisoned in 1813 for his attacks on the prince regent, later George IV. In "Poppies" from *Songs and Chorus of the Flowers*, he invested the poppy with political and philosophical significance:

> We are slumberous Poppies,
> Lords of Lethe downs,
> Some awake, and some asleep,
> Sleeping in our crowns.
> What perchance our dreams may know,
> Let our serious beauty show.

Central depth of purple,
Leaves more bright than rose,
Who shall tell what brightest thought
Out of darkness grows?
Who, through what funereal pain
Seeks to love and peace attain?

The English writer and opium addict Francis Thompson spent many years destitute and homeless. He wrote of the source of his addiction in "The Poppy":

Summer set lip to earth's bosom bare,
And left the flushed print in a poppy there:
Like a yawn of fire from the grass it came,
And the fanning wind puffed it to a flapping flame.

The color and texture of the poppy was detailed by English writer and art critic John Ruskin in *Proserpina* (1875–1886), one of his several works on the natural sciences. Chapter 4, "The Flower," explains,

We usually think of the Poppy as a coarse flower; but it is
the most transparent and delicate of all the blossoms of the
field. The rest, nearly all of them, depend on the texture of
their surface for colour. But the Poppy is painted glass; it
never glows so brightly as when the sun shines through it.
Wherever it is seen—against the light or with the light—
always, it is a flame, and warms the wind like a blown ruby.

Erasmus Darwin described the poppy in his long poem *The Loves of the Plants* (1789):

Where Sleep and Silence guard the soft abodes,
In sullen apathy Papaver nods.
Faint o'er her couch in scintillating streams
Pass the thin forms of Fancy and of Dreams.

In the language of flowers, the white poppy suggested sleep while the scarlet poppy spoke of fantastic extravagance. The oriental poppy denoted silence and the red, consolation. The church adopted the poppy for meaningful decorations: in general they represent St. Margaret, and the doubtful poppy is dedicated to the Translation of St. Edward.

PRIMROSE

This sweet Infanta of the year

The family Primulaceae includes 400 species of *Primula*. As a group of plants they are commonly called primroses, but that name specifically refers to *P. vulgaris*, the wild European primrose. *Primula* are distributed in temperate regions primarily in the Northern Hemisphere and somewhat in the tropical mountains of Java and New Guinea. Well-known species include *P. elatior*, the oxlip; *P. veris*, the cowslip, a beloved British native; and *P. auricula*, the bear's ears or auricula. The prized polyanthus primulas consist of a vast group of complex hybrids, principally of *P. vulgaris*, *P. veris*, and *P. elatior*, which come in a rainbow of colors. Others are in the Pruhonicensis Group, which is derived from *P. juliae* and *P. elatior*.

The name *Primula* is derived from *primus*, meaning first or early. The name suggests the "first rose" of the year or "prime-rose," which has been contracted into the common name primrose. The plant is considered the firstling of the year and the harbinger of not only the new life of the plant kingdom but also the renewed spirits of people and their reawakened feelings of attraction and love. The common name cowslip appears to come from the Old English *cu slippe*, for cow dung, because they are common in pastures used by cattle. The flower has such a strong positive influence that the figure of speech, "a primrose path," is still spoken of as a pleasurable, ideal course. In England, 19 April is Primrose Day, the anniversary of the death in 1881 of Prime Minister Benjamin Disraeli; the primrose was his favorite flower. Queen Victoria supplied a bouquet of primroses for his bier.

In mythology the primrose was called paralisos, which referred to the handsome son of Priapus, god of reproductive power and fertility, and Flora, goddess of flowers. Paralisos pined away and died of grief over the loss of his intended, Melicerta. He was transformed into a primrose by his parents.

Many works of literature mention primrose, cowslip, and its allied names. At one time the name primrose was applied to the daisy, *Bellis perennis*. However, by the 1500s, the name was applied exclusively to members of *Primula*.

Shakespeare knew primulas well and mentioned them numerous times. In *Hamlet* (1601) Ophelia says to her brother Laertes,

> Do not, as some ungracious pastors do,
> Show me the steep and thorny way to heaven;
> Whilst, like a puff'd and reckless Libertine,
> Himself the Primrose path of dalliance treads,
> And recks not his own rede.

In *A Midsummer Night's Dream* (1590) Oberon, king of the fairies, describes to Puck the floriferous spot in the forest where his wife Titania sleeps: "I know a bank where the wild thyme blows,/Where oxlips and the nodding violet grows." Perdita, daughter of King Leontes in *The Winter's Tale*, mentions various flowers she wishes to make into a garland for Florizel, with whom she is in love: "Pale primroses/That die unmarried, ere they can behold/Bright Phoebus in his strength." And in *Cymbeline*, Arviragus, son of King Cymbeline, agonizes over the death of his friend Fidele: "Thou shalt not lack/The flower that's like thy face, pale primrose."

An anonymous poet of the 1500s sings of sweet baziers, or bear's ears, *Primula auricula*:

Come listen awhile to what we shall say
Concerning the season, the month we call May
For the flowers they are springing, the birds they are
    singing,
And the baziers are sweet in the mornings of May.

Robert Herrick used the primrose for its connotations of first love in "The Primrose":

Ask me why I send you here
This sweet Infanta of the year?
Ask me why I send to you
This primrose, thus bepearled with dew?
I will whisper to your ears—
The sweets of love are mix'd with tears.

Izaak Walton was an English writer of the seventeenth century best known for his biographies of the literary figures of the day and for his 1653 volume on the contemplative pleasures of fishing, *The Compleat Angler.* He remarked of primroses: "When I last sat on this primrose-bank, and looked down these meadows, I thought of them as Charles the Emperor did of Florence, that they were too pleasant to be looked on but only on holidays."

It has been said that the poems of John Clare are as thickly strewn with primroses as the woodlands that he wrote about. For example, in "The Eternity of Nature" he described "Cowslips of gold bloom,/ That in the pasture and the meadow come,/Shall come when kings and empires fade and die." In "The Primrose" (1816), Clare noted,

How much thy presence beautifies the ground:
How sweet thy modest, unaffected pride
Glows on the sunny bank, and wood's warm side.
And where thy fairy flowers in groups are found.

In "Early Spring" (1860), he provided a list of the season's blooms, including the polyanthus:

> The spring is come, and spring flower coming too,
> The crocus, patty kay, the rich heartsease;
> The polyanthus peeps with blebs of dew,
> And daisy flowers; the buds swell on the trees.

The primrose was also favored by John Milton, who wrote in "On a May Morning,"

> Now the bright morning-star, day's harbinger,
> Comes dancing from the East, and leads with her
> The flowery May, who from her green lap throws
> The yellow cowslip and the pale primrose.

William Howitt noted the coincident blooming of cowslips with the emergence of new blades of grass in "Cowslips":

> Oh! Fragrant dwellers of the lea!
> When first the wild-wood rings
> With each sound of vernal minstrelsy,
> When first the green grass springs.

In *The Two Noble Kinsmen* (1613) by John Fletcher, a Jacobean dramatization of the Greek tragedy of Palamon and her soldier-lover Arcite, a boy sings the opening song as he strews flowers before a wedding procession, hailing the primrose as "first-born child of Ver,/Merry spring-time's harbinger/With her hare bells dim." And Robert Burns in "Sweet Afton" (1789) wrote,

> How pleasant thy banks and green vallies below,
> Where wild in the woodlands the primroses blow
> There oft, as mild evening weeps over the lea,
> The sweet-scented birk shades my Mary and me.

Considering the countless writers who wrote odes to the primrose or who have mentioned it in passing, the charming and graceful flower is clearly a staple of both literature and the English countryside. Appropriately, in the language of flowers it represented grace, pensiveness, and early youth.

QUEEN ANNE'S LACE

And all the meadows white

Queen Anne's lace is *Daucus carota*—the wild carrot, a member of the Umbelliferae. In England the common name refers to cow-parsley, *Anthriscus*, in the same botanical family. The Queen Anne of the common name refers not to a queen of England or France, but to St. Anne, the mother of the Virgin Mary and the patron saint of lace makers. The fine, ferny foliage and the delicate umbels of white flowers suggest a finely made lace. The genus name *Daucus* is from the Greek word *daukos*, the classical name for the wild carrot. The cultivated edible carrot is derived from the wild form and now is considered to be a distinct subspecies, *D. carota* subsp. *sativus*.

A fanciful version of the origin of Queen Anne's lace was told by Mary Leslie Newton in "Queen Anne's Lace":

> Queen Anne, Queen Anne, has washed her lace
> (She chose a summer day)
> And hung it in a grassy place
> To whiten, if it may.
>
> Queen Anne, Queen Anne, has left it there,
> And slept the dewy night;
> Then waked, to find her sunshine fair,
> And all the meadows white.
>
> Queen Anne, Queen Anne, is dead and gone
> (She died a summer's day),
> But left her lace to whiten on
> Each weed-entangled way!

RHODODENDRON & AZALEA
O rival of the rose!

With their profusion of showy flowers, at least 700 species of deciduous and evergreen woody shrubs and trees constitute the horticulturally important genus *Rhododendron* in the Ericaceae, distributed primarily in the Northern Hemisphere. The azaleas, formerly classified
in their own genus, are now included as the subgenus *Azaleastrum* within the genus *Rhododendron*. Rhododendrons and azaleas and all their
hybrids, cultivated forms, and varieties favor acid soils. Many have
rose-colored flowers, such as *R. maximum*; these are often called the
rose tree or rose bay in the southeastern United States. *Rhododendron
arboreum* is the tree rhododendron of Southeast Asia and was the first
of the Himalayan species to be introduced to Europeans in the early
1800s. *Rhododendron indicum* and its various cultivars are all azaleas. *Rhododendron prunifolium* is the plumleaf azalea, and *R. ponticum* is a species
of the Mediterranean from which many early garden varieties arose
before the twentieth-century introduction of the Asiatic species. *Rhododendron viscosum* was probably the first
North American azalea to reach
England, arriving in the 1680s.

   The name rhododendron is
from *rhodos*, for rose, and *dendron*,
for tree. It was originally applied
by the Greeks to the unrelated
shrub oleander, now *Nerium oleander*. The common name azalea is
from the Greek *azaleos*, which
means dry, a name applied because of the dry soil the plant

was erroneously presumed to prefer. Linnaeus, in his *Species Plantarum* (1753), recognized the two distinct genera *Rhododendron* and *Azalea*. His designations have not prevailed into current taxonomy because the two are botanically very similar. Many horticulturists and gardeners agree with Linnaeus, however, based on two important differences between rhododendrons and azaleas: rhododendrons have campanulate, or bell-shaped flowers, while azaleas have funnel-form flowers, and rhododendrons are largely evergreen but azaleas are generally deciduous.

In the language of flowers the rhododendron meant danger, but the azalea suggested temperance. And in the church, the rhododendron is dedicated to St. Augustine on his feast day of 26 May.

Greek soldier and essayist Xenophon narrated a history of a force of 10,000 Greek soldiers, of which he was a part, led by Cyrus in 401 BCE against his brother, the king of Persia. The somewhat embellished story tells of the soldiers being temporarily stupefied after eating honey that bees had collected from wild rhododendron (possibly *Rhododendron ponticum* or perhaps *R. luteum*) growing along the Black Sea. Such poisoning has been confirmed because the nectar does contain an affecting compound called actylandromedal.

American naturalist William Bartram traveled through the southeastern United States in the late 1700s on the first botanical survey of the Carolinas, Georgia, and Florida. He described his reaction upon seeing the flame or fiery azalea, *Rhododendron calendulaceum*, near Fort James, Georgia:

> How harmonious and sweetly murmur the purling rills and
> fleeting brooks, roving along the shadowy vales, passing
> throgh [sic] dark, subterranean caverns, or dashing over
> steep rocky precipices, their cold, humid banks condensing
> the volatile vapours, which falling coalesce in crystalline
> drops, on the leaves and elastic twigs of the aromatic shrubs
> and incarnate flowers! In these cool, sequestered, rocky
> vales, we behold the following celebrated beauties of the

hills: . . . fiery azalea, flaming on the ascending hills or wavy
surface of the gliding brooks. The epithet fiery, I annex to
this most celebrated species of azalea, as being expressive of
the appearance of its flowers, which are in general of the
colour of the finest red lead, orange and bright gold, as well
as yellow and cream colour; these various splendid colours
are not only in separate plants, but frequently all the varieties and shades are seen in separate branches on the same
plant; and the clusters of the blossoms cover the shrubs in
such increasing profusion on the hillsides, that suddenly
opening to view from dark shades, we are alarmed with the
apprehension of the hill being set on fire. This is certainly
the most gay and brilliant flowering shrub yet known.

Ralph Waldo Emerson put the beauty of a woodland rhododendron into the first rank of all flowers in "The Rhodora" (1834):

On being asked, whence is the flower?

In May, when sea-winds pierced our solitudes,
I found the fresh Rhodora in the woods,
Spreading its leafless blooms in a damp nook,
To please the desert and the sluggish brook.
The purple petals, fallen in the pool,
Made the black water with their beauty gay;
Here might the red-bird come his plumes to cool,
And court the flower that cheapens his array.
Rhodora! If the sages ask thee why
This charm is wasted on the earth and sky,
Tell them, dear, that if eyes were made for seeing,
Then Beauty is its own excuse for being:
Why thou wert there, O rival of the rose!
I never thought to ask, I never knew;
But, in my simple ignorance, suppose

> The self-same Power that brought me there brought
>   you.

William Howitt wrote of the mountain rhododendron, possibly *R. fer-rugineum*, in "The Rhododendron on the Alps":

> And is it here, that sunny flower
> That decks our gardens so?
> And can it brave the mountain storm
> Where the oak cannot grow?
> 'Tis joy upon our frozen way,
> Amid the Alpine gloom,
> Amid the glaciers and the snows,
> To see its crimson bloom.

ROSE

The flower most holden in prize

The genus *Rosa* consists of about 150 rambling or erect mostly deciduous and a few evergreen shrubs. The native distribution of *Rosa* is
primarily in the temperate regions of the Northern Hemisphere and
in the mountains of the tropics. It is the namesake of the vast Rosa-
ceae, a family full of well-known and widely cultivated ornamental
plants, including fruit trees, whose members often have thorns, prick-
les, or bristles. The familiar shrubs of the genus *Rosa* bear flowers with
five petals and numerous stamens. The rose is a universally adored
flower that symbolizes love and ro-
mance as well as war and peace.
It has been used in pagan and
secular rites and in Christian
ceremonies. The flower has
many associations with English
history; most recently the white
rose has become the symbol of
the late Diana, Princess of Wales.

Thousands of roses are grown
and valued as cut flowers and for
the scent of their essential oils.
Many cultivars have double sets
of petals. *Rosa banksiae*, Banksia rose;
R. chinensis, China rose; *R. ×damascena*,
damask rose; and *R. rubiginosa*, the
eglantine or sweet brier, are well
known. One of the oldest cultivated
roses is the apothecary rose, *R. gallica* var.

officinalis, which was the emblem of the House of Lancaster during the Wars of the Roses from 1455 to 1485. The House of York, the other warring faction, chose *R. ×alba*, or the white rose of York, as its heraldic symbol. The war ended with the establishment of the House of Tudor whose emblem was the Tudor rose, *R. ×damascena* 'Versicolor', a double rose usually depicted with an outer circle of red petals and an inner of white.

The name *rosa* is the classical Latin name for the plant, literally translating as red, the most common color of the flower. In classical Greek, the rose appeared as *rhodon*, from which are derived the names rhododendron, literally rose trees, and Rhodes or Rodos, the Greek island where roses flourish. The word *rosa* is also related to rosary beads, which according to church doctrine represent the crown of thorns, or *rosarium*, worn by the Virgin Mary. The rose is closely identified with Mary in church iconography; a standard component of gothic cathedrals was a rose window, a large, circular window of stained glass, through which Mary's visage shone. Mary is also known as the Rose of Sharon. In literature, biblical writing, lore, and horticulture, the rose of Sharon is a popular name appropriated by various plants, including hibiscus, St. John's wort (*Hypericum*), narcissus, tulip, and saffron crocus, to name a few. These various "roses" of Sharon are so called because of their ability to survive on the Plain of Sharon, a region in Israel along the Mediterranean Sea difficult to cultivate because of its soil conditions.

According to one myth, the term *sub rosa*, which suggests strict confidence or private conversation, originated when Cupid used a rose to bribe Harpocrates, the God of Silence, into keeping quiet about Aphrodite's various love affairs. Harpocrates is usually depicted as a handsome young man holding a white rose in one hand, with a finger of the other hand at his lips. Hence came the saying "under the rose," which was used to mark secrets. Roses were hung over banquet tables by the Romans and later were designed into the ornamental motif of the ceiling—decorations that were appropriately called roses or rosettes—to remind guests not to repeat conversations heard at the table. Contemporary formal houses often include such a rosette from which a chandelier hangs over the dining table.

Aphrodite, the Greek goddess of love and beauty, Venus to the Romans, was the source of the red rose. When she pricked herself on the thorn of a white rose, the blood from her finger dripped on its petals and stained them forever. The rose is sometimes depicted in the crown of Venus or on a scepter that she carries. Other legends claim the red petals resulted when Cupid spilled red wine on the petals of a white rose. In ancient times roses were the customary flower to weave into chaplets for the crowns of brides and bridegrooms.

In the language of flowers, roses suggested love and beauty. The damask rose indicated a brilliant complexion, the musk rose suggested capriciousness, and the red rosebud said, "You are young and beautiful." In the church, white and red roses are dedicated to St. George on his feast day of 23 April; the blush rose is offered to St. Mary Magdalene on 22 July; the yellow rose is for St. Nicomede on 1 June; and the three-leaved rose is dedicated to St. Boniface on 5 June.

On festive occasions in ancient Rome, roses were featured on the garlands worn by Roman officials. The blooms were also strewn along the streets and often placed decoratively upon columns and public statues. Such customs were described in ancient Latin texts. Horace, in his lyric poem *Odes*, said about the celebration of the return of a close friend, "Nor let roses be wanting to our feast." In Lucius Apuleius's *Metamorphoses*, the only Latin novel that survives in its entirety, Venus is described after an evening of revelry "Heavy with wine and all her body bound about with flashing roses."

Before the 1500s few roses grew in Europe. In the 1700s, however, with the arrival from China of long-blooming species, including the sweetly scented tea roses, the stage was set for new hybrid tea roses, the first of which was *Rosa* 'La France', a descendant of the hybrid perpetual roses.

Western literature abounds with references to roses. Chaucer described Cupid clad in roses in *Romaunt of the Rose*: "And many a rose-leaf ful long/Was entermedled there-among:/And also on his head was set." *The Song of Roland*, an anonymously written Old French epic dating from perhaps as early as the mid-eleventh century, mentions the eglantine rose, *Rosa rubiginosa* (translated by C. S. Moncrieff):

> On white carpets those knights have sate them down,
> At the game-boards to pass an idle hour—
> Chequers the old, for wisdom most renowned,
> While fence the young and lusty bachelours
> Beneath a pine in eglantine embow'red.

"All Night by the Rose," an anonymous work from the thirteenth century, describes a man's encounter with a young virgin, here represented by a rose:

> All night by the rose, rose—
> All night by the rose I lay;
> Dared I not the rose steal,
> And yet I bore the flower away.

In English ballads it was common for young lovers to be buried side by side, and oftentimes a rose and a sweet briar would grow from their graves, as is the case in "Margaret and William."

> Margaret was buryed in the lower chancel,
> And William in the higher,
> Out of her brest there sprang a rose,
> And out of his a briar.
> They grew till they grew up to the church top,
> And then they could grow no higher;
> And there they tied a true lover's knot
> Which made all the people admire.

Shakespeare mentioned roses at least sixty times in his works. One of his sonnets (number fifty-four) cites the canker or dog rose, *Rosa canina*, an Old World native. In addition to using the legitimate name canker rose, Shakespeare also raises the image of petals diseased by canker, a word from the Old English and Latin *cancer*:

> The rose looks fair, but fairer we it deem
> For that sweet odour which doth in it live.
> The canker-blooms have full as deep a dye
> As the perfumed tincture of the roses.

Well-known poets such as Milton, Shelley, Spenser, Burns, and others wrote rapturously of the rose and its many symbolic connotations. Less well-known English naturalist Richard Jefferies also was captivated by its spell. In *Wild Life in a Southern County* (1879) he described a rose hedge near a farmhouse at Wick in Wiltshire that provided curative powers:

> The great meadow hedge—the highway of the birds—
> where it approaches the ha-ha wall of the orchard, is lovely
> in June with the wild roses blooming on the briars which
> grow there in profusion. Some of these briars stretch forth
> into the meadow, and then, bent down by their own weight,
> form an arch crowned with flowers. There is an old supersti-
> tion about these arches of briar hung out along the hedge-
> row: magical cures of hooping-cough and some other dis-
> ease of childhood can, it is believed, be effected by passing
> the child at sunrise under the briar facing the rising sun.

But Jefferies added, "This has to be performed by the 'wise woman' in the hamlet, who retains a reputation for witch-craft."

Poet Tadhg Dall O'Huiginn, composer of about forty poems, was blind most of his life. Eventually he was murdered by one of the O'Hara clan, whom he had satirized in a poem. In "The First Vision," translated by the Earl of Longford, the poet dreamed of seeing the queen of fairyland:

Fair was her face, her cheeks outblushed the rose;
There might you see the floods of crimson rise,
And dark unfaltering brows above disclose
The hyacinthine petals of her eyes.

A Christmas carol dating from the 1400s describes the rose within an account of the spread of Christianity to France:

The rose is the fairest flower of all,
That evermore was, or evermore shall,
The rose of ryse.

. . . .

The rose is the flower most holden in prize.
Therefore, me thinks, the fleur-de-lys
Should worship the rose of ryse
And be his thrall.

And religious reasons are used to explain the origin of the rose's thorns in "The Rose" from *His Noble Number* (1648) by Robert Herrick:

Before man's fall the rose was born,
St. Ambrose says, without the thorn:
But for man's fault then was the thorn
Without the fragrant rose-bud born;
But ne'er the rose without the thorn.

Probably the most famous of lines written about roses was penned by Herrick in "To the Virgins, to Make Much of Time":

Gather ye rosebuds while ye may,
Old Time is still a-flying:
And this same flower that smiles to-day
To-morrow will be dying.

And Thomas Hood recalled vividly the season he met his love in "Time of Roses":

> It was not in the Winter
> Our loving lot was cast;
> It was the time of roses—
> We pluck'd them as we pass'd.

Finally, perhaps the simplest of all invocations to the rose is paid in this couplet by Ebenezer Elliott titled "On a Rose in December" (1876):

> Stay yet, pale flower, though coming storms will tear
>   thee,
> My soul grows darker, and I cannot spare thee.

SAINT JOHN'S WORT

Bloom here, thou plant of power

A handful of species of *Hypericum* are called St. John's wort. The whole genus consists of about 400 generally yellow-flowered species of trees, small shrubs, and herbs that are distributed worldwide. *Hypericum* is a member of the Guttiferae or mangosteen family. The St. John's wort of legends and plant lore is *H. perforatum*, although the ancients knew and probably used in ceremonies *H. olympicum*, *H. trichocaulon*, and *H. linarioides*—the last two names were formerly misapplied to *H. reptans*.

In North America, St. Andrew's cross, *H. hypericoides*, was recorded as early as the eighteenth century by William Byrd of Virginia, who noted that the natives used the plant to cure snakebite.

The Greeks gave the name hypericum to the plant that was placed above —*hyper*—doorways and religious figures—*eikons* for icons—to ward off evil during the midsummer festival on the eve of the longest day and shortest night of the year, usually about 21 June. Saint John's wort was known as *sol-terrestris*, or terrestrial sun, because of its bright yellow midsummer-blooming flowers. The flowers symbolized St. John, who was a "light to them which sit in darkness." Midsummer's eve was the occasion for the feast day of St. John the Baptist (which is now set as 24 June); bonfires were lit, lamps and candles burned throughout the night, and wreaths of hypericum hung around the houses and doorways to keep out the devil—the symbol of darkness and evil—and phantoms, specters, even storms and lightning.

The leaves of *Hypericum perforatum* appear to be pierced by innumerable holes because of translucent glands; these are easily visible when held to the light. According to tradition, the devil pricks or perforates the plant to diminish its powers. In French, the hypericum is called *millepertuis*, meaning "a thousand holes." Legend claims that the spots turn red and ooze blood on 29 August, the date commemorating St. John's beheading.

Hypericum also was called *fuga daemonum*, or "The Flight of Devils," by medieval writers. An anonymous poem explains this superstition: "St. John's Wort, scaring from the midnight heath/The witch and goblin with its spicy breath." In the Middle Ages, so powerful was the belief in the effectiveness of the plant's general curative powers that it was known as *tutsan* or *la toute-saine*, meaning all saint or all heal. But in 1646 Sir Thomas Browne, explaining false opinions about "the endeavors of satan," wrote in book 1, chapter 10 of *Pseudodoxia Epidemica* "That any virture there is in Hipericon to make good the name of Fuga Daemonis . . . is not easy to believe."

Robert Burton in *Anatomy of Melancholy* (1621) said the plant cures melancholy: "Bassardus Viscentius . . . commends Hypericon, or St. John's wort gathered on a Friday in the hour of Jupiter, when it comes to his effectual operation (that is about the full moon in July); so gathered and borne, or hung about the neck, it mightily helps this affection, and drives away all phantastical spirits." William Cowper, an eighteenth-century hypochondriac who likely used St. John's wort to treat depression, wrote in "The Winter Walk at Noon" about the plant that would bloom the following summer: "Hypericum all bloom, so thick a swarm/Of flowers, like flies, clothing its slender rods,/That scarce a leaf appears."

Saint John's wort was used as a prognosticator of marriages. In Saxony, young girls placed hypericum over their beds at night; if the plant remained fresh through the night it foretold marriage within a year. An anonymous poet wrote of a negative prophesy from the plant:

> The young maid stole through the cottage door,
> And blushed as she sought the plant of power;
> "Thou silver glowworm, O lend me thy light,
> I must gather the mystic St. John's-wort to-night;
> The wonderful herb, whose leaf will decide
> If the coming year shall make me a bride!"
> And the glowworm came
> With its silvery flame,
> And sparkled and shone
> Thro' the night of St. John;
> And soon as the young maid her love-knot tied,
> With noiseless tread
> To her chamber she sped,
> Where the spectral moon her white beams shed.
> "Bloom here, bloom here, thou plant of power,
> To deck the young bride in her bridal hour?"
> But it drooped its head, that plant of power,
> And died the mute death of the voiceless flower;
> And a withered wreath on the ground it lay,
> More meet for a burial than bridal day.
> And when a year was past away,
> All pale on her bier the young maid lay!
> And the glowworm came
> With its silvery flame,
> And sparkled and shone
> Thro' the night of St. John;
> And they closed the cold grave o'er the maid's cold clay.

The Scottish have an old name for hypericum, *ach larson cholumcille*, or "the little arm pit of Columba." In the Scottish isles, the plant is dedicated to St. Columba. The saint once found a child, whom he had engaged to herd his cattle, weeping because he was afraid of the dark and worried that the cattle might stray away. Saint Columba plucked St. John's wort and put it under the child's arm, telling him to sleep in peace for no harm would come to him or his charges.

Despite the overwhelming powers for good attributed to St. John's wort, the language of flowers considered it a suggestion of animosity.

SALVIA

Its smell is somewhat vehement

The 900 or so species of *Salvia* are perennial shrubs and herbs of the Labiatae, the mint family. Although they are found in most parts of the world, about half occur in the Americas, primarily in subtropical regions, and none are from Australia or the islands of the southwestern Pacific. Salvias, or sages, are highly valued border and bedding plants. The common aromatic culinary sage is *S. officinalis*. Clary is the biennial *S. sclarea*, whose distillation yields a muscat scent used in perfumes and wines. The flower heads of clary sage yield a resin called sclareal, an important flavor for tobacco and a starting material for the perfume, flavor, and fragrance industries. Combining the distillation with brandy and cinnamon produces a cordial called clary-water.

American salvias began appearing in Europe in the late 1700s. One of the earliest was *Salvia coccinea*, found by John Bartram, the father of William Bartram, on his travels through Florida in 1765. *Salvia guaranitica*, native to Brazil and Argentina, has a deep blue flower; one pale blue form is known as 'Argentine Skies'. *Salvia splendens* is the scarlet sage popularly used as a summer bedding plant. The beautiful design on the seven-branched menorah is said to have been inspired by *S. judaica*, a sage native to Palestine.

The name salvia is from the Latin *salvus*, which means safe or heal or "to be in good health." The saving herb, as it is sometimes known, refers to the manifold medicinal properties of the plant, used particularly in medieval times. The specific epithet *officinalis* of the highly esteemed common sage means "of the apothecary," which suggests that it was generally used as a home remedy or domestic cure. The common name, sage, is probably derived phonologically from the French *sauge*—both names refer specifically to *S. officinalis*. So touted were its numerous properties that an old proverb asked, "*Cur morietur homo cui Salvia crescit in horto?*" (How can a man die who grows sage in his garden?) Another old saying from the British Isles claimed, "Sage helps the nerves and by its powerful might palsy is cured and fever put to flight." Sage would mitigate chills and keep away toads, and as John Gerard said, "Being put up into the nosthrils, it draweth thin flegme out of the head." According to legend such amazing healing power came from the salvia because it had been blessed by the Virgin Mary; during the flight from Egypt a sage plant gave her shelter from Herod's men.

Salvias are well known in myth and folklore. In one tale, a young, shy nymph lived as the sage plant in a hollow oak. Below her tree a group of garish jonquils grew and flourished, diminishing the beauty of the reticent Sage. She was not jealous and continued to live alone shyly in the quiet woods until one day a king arrived with his huntsmen and hounds. Sage was immediately charmed by the king but knew that it would be death for her to love a mortal. Nevertheless, she was smitten by him, and he by her. She told him that the best days were gone, but the solitude of the woods was still beautiful, so they should remain there. "You ask my love and I give you my life," she told him. But the king did not understand what she meant as he folded her passionately into his arms. She returned his caresses, paled, and her head drooped. He tried to revive her with water from a nearby pool, but the heat of their passion was more than the fragile Sage could endure, and life ebbed from her. The king went away confused and mourned her passing.

According to another legend, young girls who pick sage at midnight on Christmas Eve, or on midsummer's night in other versions, will behold the image of their future husband.

Salvias can be fragrant or even acrid, depending on the species. The gardener and poet Walafrid Strabo in *Hortulus* considered their fragrance sweet, but their disposition ruthless:

> In the front side shines the sweet smelling sage,
> In perpetual youth it grows forever green.
> Because it is good to mix and use and has virtues
> That aid many ailments.
> But withal is the seed of uncivilness
> Because unless the new shoots are cut off, it turns
> Wildly on its forebears and kills them—
> The old roots and stem in jealousy.

But John Gerard noted a pungent odor in his description of the plants:

> The stalkes are rough and hairie, four square below, and round at their tops. The floures in their growing and shape are like those of the ordinarie, but of a whitish purple colour, and fading, they are each of them succeeded by three or foure seedes, which are larger than in other sages, and to fill their seed-vessels, that they shew like berries. The smell of the whole plant is somewhat more vehement than that of the ordinarie; the leaves are sometimes little eares or appendices, as in the smaller of Pig-Sage.

In the language of flowers, the common sage suggested domestic virtue and the garden sage meant esteem. The blue salvia conveyed the sentiment, "I think of you," whereas the red salvia spoke, "Forever thine."

SEDUM

That bright weed, the yellow stone crop

About 300 typically succulent species of *Sedum* are distributed through-out the Northern Hemisphere, including mountainous regions of the tropics. Sedums, commonly called stone crops, are members of the Crassulaceae and have simple fleshy leaves. The brightly yellow-flow-ered *S. acre* is a widely grown rampant mat-forming species.

Sedum is the classical Latin name applied to succulent plants of var-ious genera and is derived from the Latin *sedeo*, for "I sit," an allusion to the manner in which the plants sit among rocks and cover them. The common name stone crop has been applied because of the grayish, stony ap-pearance of the plants and their suggestion of a crop of stones growing among boulders and rocks. Sedums are sometimes confused with houseleeks or orpines, members of the genus *Sem-pervivum*, which is also in the Crassulaceae. Older usage often interchanged their common names. The stone crop was referred to in old herbals as English mouse tail, prick madam, wall pepper, and poor man's pepper; the last arose from the bitter taste of the leaves of sedum, an "herb of a very hot tem-perature, sharp and biting."

John Gerard described the stone crop or wall-pepper in 1633, mixing houseleeks or prick-madams with sedums:

> This is a low and little herbe; the
> stalks be slender and short; the leaves

about these stande very thicke, and small in growth, full
bodied, sharped pointed, and full of juyce. The flowers
stand of the top, and are maruelvous little, of colour yellow,
and of a sharpe biting taste. The root is nothing but strings.
It groweth everywhere in stony and dry places, and in chinks
and crannies of old wals, and on the tops of houses. It is alwaies green, and therefore it is very fitly placed among the
sengreenes.

In the language of flowers the stone crop suggested tranquillity,
which relates to another belief that growing it on the roof or around
the house would ward off lightning. In flowers appropriate for church
decorations, the great stone crop, possibly *Sedum spectabile*, now *Hylotele-
phium spectabile*, is used on the feast day of St. Giles on 1 September.

John Clare, though frequently confined to a mental asylum, in his
poetry remembered happier times at home with his family. He mentioned the houseleek in his reminiscences, but he was probably referring to a stone crop, perhaps *Sedum reflexum* or *S. acre*, both western European natives:

> How oft I've stood to see the chimney pour
> Thick clouds of smoke in columns lightly blue,
> And close beneath the house-leek's yellow flower,
> Approaching to a nearer view.

London's Blue Stocking Circle of the late eighteenth century derived
its name from the blue worsted stockings that the intelligent and
learned women who formed the group wore to its gatherings. One
member of the circle, Hannah More, wrote poems, a novel, and numerous essays. In "The Bleeding Heart," she described the tradition of
maidens carrying home stone crop flowers on midsummer's eve; hanging them would determine the faithfulness of their lovers. If during
the night the flower turned to the right the lover was faithful; if to the
left, he was untrue.

Now once a year, so country records tell,
When o'er the heath sounds out the midnight bell,
The Eve of Midsummer, that foe to sleep,
What time young maids their annual vigil keep,
The tell-tale shrub fresh gathered to declare,
The swains who false from those that faithful are.

"The Excursion" (1814) and "The Recluse" (1806) by William Wordsworth are serial poems composed in a didactic, narrative style. Though
written later, "The Excursion" tells the first part of the story of Margaret, a woman who once lived in a ruined, overgrown cottage:

The honeysuckle, crowding round the porch,
Hung down in heavier tufts, and that bright weed,
The yellow stonecrop, suffered to take root,
Along the window's edge, profusely grew,
Blinding the lower panes.

George Manville Fenn made reference to the softness of the sedum
leaves when he wrote *In An Alpine Valley* in 1894: "He threw himself
down upon some bed of sedums, where quite a couch was formed of
the tiny rosettes."

Walt Whitman reflected on the stonecrop in "Song of Myself"
(1855), a monumental work in which he gave everything significance,
celebrating himself, all things, and all people:

I accept Reality and dare not question it,
Materialism first and last imbuing.
Hurrah for positive science! long live exact demonstration!
Fetch stonecrop mixed with cedar and branches of lilac,
This is the lexicographer, this is the chemist, this made a
    grammar of the old cartouches,
These mariners put the ship through dangerous unknown seas,
This is the geologist, this works with the scalpel, and this is
    a mathematician.

SNAPDRAGON

The stern and furious lion's gapping mouth

The much-loved snapdragons of gardens are the descendants and cultivars of *Antirrhinum majus*, a member of the Scrophulariaceae or foxglove family. *Antirrhinum majus* is typically treated as an annual bedding plant. Snapdragons and more than forty sibling species grow naturally in most temperate regions of the Northern Hemisphere. Annual or perennial, they have characteristically two-lipped flowers, usually borne on erect terminal racemes. The garden snapdragon, which originated in the Mediterranean, has many forms: some are dwarf, such as *A. majus* 'Tom Thumb', and others are tall, such as the rocket and the penstemon-like types. One species from Spain, *A. hispanicum*, has a dwarf shrubby habit, and a species from the American Southwest, *A. multiflorum*, has pink to carmine flowers.

The name antirrhinum is from Greek *anti*, for opposite, and *rhis*, for snout or nose. Both the Latin and common names, then, allude to the flowers' shape, which is vaguely reminiscent of a dragon's snout. Besides snapdragon, former common rural English names include lion's snap, toad's mouth, calf's snout, and dog's mouth. The Roman writer Columella called it "the stern and furious lion's gapping mouth." Such common names make sense for the delicate bloom if one pinches the throat of the flower and witnesses, though in miniature, the apparent gapping maw.

In *Paradisi in Sole, Paradisus Terrestris* (1629), John Parkinson wrote of the snapdragon, "There is some diversity in the snapdragons, some being of a larger, and others of a lesser stature and bignesse; and of the larger, some of one, and some of another colour, but because the small kindes are of no beautie, I shall at this time onely entreate of the greater sorts." Of its "vertures," Parkinson said, "They are seldome or never used in physicke by any in our dayes."

Snapdragons are supposed to have supernatural powers, to be able to destroy charms, and to protect from witchcraft. They also provide shelter for elves. A German legend tells of a housewife who was kidnapped by an elf. While leading the woman away from her home, the elf told her to lift the hem of her skirt to avoid stepping on the origanums and the snapdragons. The astute woman immediately stepped on the snapdragons, crushing them and destroying their ability to safeguard the elf, from whom she promptly escaped.

Oil can be extracted from the seed of snapdragons, and great fields of them are said to have been grown for this purpose in Russia. How the oil came to be discovered is told in a Russian tale in which a poor woodcutter lived on a small plot of land in the midst of vast fields of snapdragons. One day as he was going home to have lunch, a small man stopped him and asked if he had a bite of food to share because the man had traveled a great distance without anything to eat. The kind-hearted woodcutter invited the man in and shared the last of his black bread. When the woodcutter apologized for having no butter to spread on the bread, the little man said, "Wait. I have a solution." He dashed outside and brought in heads of ripened snapdragons. He crushed the seed, and oil oozed onto the bread. After eating his fill, the little man thanked the woodcutter for sharing his bread and went on his way. The woodcutter began expressing oil from the abundant snapdragon seed all around him and sold it in the nearby village, greatly improving his financial lot.

Metaphysical poet and essayist Abraham Cowley wrote of the snapdragon's magical powers in book 4 of his *History of Plants*:

Antirrhinum, more modest, takes the style
Of Lion's-mouth, sometimes of Calf's-snout vile,
By us Snapdragon called, to make amends,
But say what this chimera name intends?
Thou well deserv'st it, as old wives say,
Thou driv'st nocturnal ghosts and sprights away.

In the language of flowers the snapdragon conveyed the suggestion of presumption, because the flower presumes to look like an animal. Medieval legends state that maidens who wear the flowers will appear gracious in the sight of others.

SNOWDROP

Pretty firstling of the year

About fifteen recognized species of *Galanthus*, the snowdrop, are in the family Amaryllidaceae, including many named garden varieties. The original geographical distribution of the snowdrop is from western Europe to the Caucasus Mountains of Iran and the Caspian Sea. Well known is *G. elwesii*, or the giant snowdrop, from Turkey, but the smaller common snowdrop, *G. nivalis*, is the most widely grown in gardens of North America and England. *Galanthus reginae-olgae* is an autumn-flowering snowdrop from the Balkans, and *G. plicatus* was originally collected and returned to the British Isles during the Crimean War by Lord Clarina and others. Snowdrops as a group of plants are easy to recognize, but the garden forms are difficult to sort.

Carl Linnaeus derived the common snowdrop's Latin name, *Galanthus nivalis*, from the Greek *gala*, for milk, and *anthos*, for flower, and the species name *nivalis* means "of the snow"; hence, its full name rather poetically translates as "milk flower of the snow." *Galanthus elwesii* is named for English naturalist H. J. Elwes, who introduced the species into cultivation; *plicatus* refers to the folded pleats of the leaves; and *reginae-olgae* to honor Queen Olga of Greece.

The English name snowdrop does not refer to a drop of snow but rather to white, drop-shaped pendants worn as earrings and called drops or eardrops. They appeared on women in Dutch and Italian paintings in the fifteenth and sixteenth centuries. Because the earrings were so common they were associated with the white flowers of snow-bells, which gradually came to be called snowdrops. In Holland the flowers are also known as foolish maids, and in other countries milk flower and fair maids of February. In French the snowdrop is belle of the snow, winter bell, and *perce-neige*, which literally translates as snow-piercer. In German it is called snow violet and *Schneetropfen* (snowdrop).

The single day in the church calendar that has been traditionally associated with the snowdrop is 2 February, Candlemas Day. In the Catholic, Greek, and Anglican churches, Candlemas is the day dedicated to the Purification of the Virgin Mary and occurs forty days after Christmas. The purification ritual for mothers was held according to Mosaic law (Lev 12:1–8) forty days after the birth of a male child. Anything considered impure was removed through the ritual; Mary made an offering of a lamb and a pigeon to symbolize her purity. The snowdrop's religious association has led to other common names, including Candlemas bells, purification flower, Candlemas flower, and Mary's tapers. Historically Candlemas is the day to remove from the church the Christmas greenery and the various icons and images of the Madonna. Snowdrops plucked from convent gardens replaced the Christmas decorations and were strewn about the altars as emblems of the Virgin's chastity and purity. Legends claimed that if the Christmas greenery were not removed, goblins would appear and fill the church in numbers equal to the number of remaining leaves.

Candlemas is also the celebration of the Feast of the Presentation of Christ in the temple in eastern Christian churches. Since the eleventh century, candles have been lit and blessed on Candlemas Day because they symbolically represent the biblical story in which Simeon, the devout old priest, told Mary that Jesus would be the light of the world: "a light to lighten the Gentiles, and the glory of thy people Israel" (Luke 2:32). At the end of Mary's trip from Bethlehem to Jerusalem she carried the infant Jesus to a temple to be presented; legend describes chaste snowdrops springing from her footprints.

An often-cited poem on "flower garlands for the seasons" begins the garden year with a reference to snowdrops, explaining their time of bloom and that of other early flowers. It is from "An Early Church Calendar of English Flowers" which probably dates from the 1500s:

> The snowdrop, in purest white arraie,
> First rears her hedde on Candlemas daie;
> While the crocus hastens to the shrine
> Of Primrose love on St. Valentine.

The second day of February also has long been a day for providing seasonal weather predictions because of the traditional belief that hibernating animals wake up on this day to see if it is still winter. Groundhogs, or woodchucks, badgers, hedgehogs, bears—and snowdrops—have been meteorological prognosticators in Europe and North America. In the United States Candlemas is better known as Groundhog Day.

A legend explains the tapered petals of the snowdrop and their winter bloom. When Adam and Eve were banished from the Garden of Eden, Eve was despondent and lonely because the world was dark, cold, and barren. Summer had turned to winter, and driving snow cast a pall on the earth. An angel, seeing Eve, took pity on her and breathed on the falling flakes of snow, turning them into snowdrop flowers. Tapered and bowing humbly at Eve's feet, each flower revealed a touch of green on its petals in a promise that the cold winter would soon end and summer would return. The story was told by an anonymous poet:

The angel's visit being ended,
Up to Heaven he flew;
But where he first descended,
And where he bade the earth adieu,
A ring of Snowdrops formed a posy
Of pallid flowers, whose leaves, unrosy,
Waved like a winged argosy,
Whose climbing mast above the sea,
Spread fluttering sail and streamer free.

And an anonymous quatrain restates the snowdrops' promise of spring:

And thus the snowdrop, like the bow
That spans the cloudy sky,
Became a symbol, whence, we know
That brighter days are nigh.

Erasmus Darwin, the grandfather of Charles, was a physician, botanist, poet, and thinker. He wrote two long didactic poems, *The Economy of Vegetation* and *The Loves of the Plants*, in which the goddess of botany comes to earth to explain the Linnaean system of natural classification. His works include embryonic theories of evolution (although not the same as those of his yet-unborn grandson). The poems were published together in 1791 as one volume called *The Botanic Garden*. In this excerpt from *The Loves of the Plants*, Galantha, the snowdrop goddess, calls the spring to awake:

Warm with sweet blushes bright Galantha glows
And prints with frolic step the melting snows:
O'er silent floods, white hills, and glittering meads,
Six rival swains the playful beauty leads,
Chides with her dulcet voice the tardy Spring,
Bids slumbering Zephyr stretch his folded wing,

Wakes the hoarse Cuckoo in his gloomy cave,
And calls the wondering Dormouse from his grave,
Bids the mute Redbreast cheer the budding grove
And plaintive Ringdove tune her notes to love.

The snowdrop also has a dark side in some legends. Though it is acceptable to pick a bouquet of snowdrops, a single first flower of the season is called a death token if it is picked and brought into the house; death will come to someone in the family if the first snowdrop is picked. A sign of this curse is in the shape of the flower, which resembles the shroud of a corpse. In fact, it is said to be unlucky to bring into the house a single first flower of primrose, violet, daffodil, or any other spring blossom.

The Danish writer Hans Christian Andersen collected and wrote fairy tales and other children's short stories in the nineteenth century. Among his works is "The Snowdrop" (1863), which, along with the flower, is called *Sommer Gorrk* (Summer Fool) in Danish. In the tale, an alluring and enticing winter sunbeam shone on a little snowdrop while some children were playing around it. Together they charmed the tender plant by admiring its dress and inviting it to come out to play. When it did, a little boy picked it and took it inside his house. There, the oldest daughter in the family took it and sent it as an April fool's joke to a boyfriend in a distant town. The boyfriend was to be her short-term "summer fool," for she designed to find another boyfriend by midsummer, when the snowdrop would be long forgotten.

In the Northern Hemisphere snowdrops are the traditional midwinter floral heralds, often poking their solitary, pendent, white drops through layers of snow by early January when few other bulbous plants have broken dormancy. Their offering of hope and consolation was restated in the language of flowers. Many writers have made the snowdrop a symbol of renewal and the return of better times. Englishman Thomas Tickell wrote in "Kensington Garden,"

A flower that first in this sweet garden smil'd,
To virgins sacred, and the snow drop styl'd
The new-born plant with sweet regret she view'd
Warm'd with her sighs, and with her tears bedew'd

. . . .

Mid frosts and snows triumphant dares appear,
Mingles the seasons, and leads the year.

Scottish-born poet James Thomson, a forerunner of the romantic style, wrote a collection of popular poems on the natural stages of the year called *The Seasons* (1726–1730). In "Spring," he wrote, "Fair-handed Spring . . ./Throws out the snowdrop, and the crocus first." Brian Waller Proctor, a London barrister, wrote "To the Snowdrop" under the pseudonym Barry Cornwall:

Pretty firstling of the year!
Herald of the host of flowers!
Hast thou left thy cavern drear,
In the hope of summer hours?

American Celia Thaxter gardened off the coast of Portsmouth, New Hampshire, on Appledore, one of the Isles of Shoals. Her book, *An Island Garden* (1894), describes the snowdrops heralding springtime:

While I am busy with pleasant preparation and larger hope, I rejoice in the beauty of the pure white Snowdrops I found blossoming in their sunny corner when I arrived on the first of April, fragile winged things with their delicate sea-green markings and fresh, grass-like leaves. Ever since the first of March have they been blossoming, and the Crocus flowers begin, as if blown out of the earth, like long, lovely bubbles of gold and purple, or white, pure or streaked with lilac, to break under the noon sun, into beautiful petals showing the orange anthers like flame within.

The snow-white color of the snowdrop has struck other writers. William Wordsworth noted in "To a Snowdrop," "Lone flower, hemmed in with snows, and white as they,/But hardier far, once more I see thee bend/Thy forehead, as if fearful to offend, Like an unbidden guest." Vicar Francis Kilvert, at St. Harmond near the Welsh border in England, wrote in his diary on 3 March 1878: "As I walked in the Churchyard this morning the fresh sweet sunny air was full of the singing birds and the brightness and gladness of the Spring. Some of the graves were as white as snow with snowdrops." Expressing a somewhat contrary opinion, Tennyson wrote in "The Last Tournament" (1871) from his *Idylls of the King*, "The snowdrop only, flowering through the year,/Would make the world as blank as winter-tide." These lines are spoken to Tristram by a damsel whose "white day of Innocence hath past."

A practical use for snowdrops is noted by Margaret Plues in *Rambles in Search of Wild Flowers and How to Distinguish Them* (1879):

> Everyone hails the Snowdrop with delight. The weary winter is passed, and its stern hand has bowed down many a beloved head whose hope of health is dependent on the coming of spring. A few of these pale blossoms carried to the sick bed, or placed beside the worn invalid, bring with them fresh hopes of new-sprung courage. Children rejoice in the white buds, and in their rapid sanguine nature feel that hosts of bright flowers will be open on the morrow, for that the spring has come indeed.

Samuel Taylor Coleridge wrote the following in "The Snowdrop":

> Fear no more, thou timid Flower!
> Fear no more the winter's might,
> The whelming thaw, the ponderous shower,
> The silence of the freezing night.

But despite the frail and tender look of the snowdrop, it is extremely hardy. Pushing indomitably from amid layers of dead leaves or snow during the coldest, bleakest days of winter, it is truly an emblem of gardening hope. The snowdrop is the reliable voluntary to the sweet floral music of spring.

SPEEDWELL

With their bright blue eyes

Speedwell, or *Veronica*, is a member of the Scrophulariaceae, the figwort family. The 250 species of *Veronica* are native to the northern temperate regions, particularly Europe and Asia Minor. They are primarily perennial herbaceous plants, many of which are grown as ornamentals. Widely grown are the many cultivars of *V. longifolia* and *V. spicata* as well as the sprawling *V. prostrata*. The germander speedwell, *V. chamaedrys*, has vivid blue flowers and goes by the appellations angels' eyes, bird's eye, and God's eye.

The name veronica is a compound of *vera* and *icon* from both Latin and Greek. Meaning "true image," the name is associated with St. Veronica, whose handkerchief miraculously recorded the image of Christ's face. According to legend, she wiped his brow with the handkerchief while Christ was en route to his crucifixion on Calvary. The handkerchief became a holy relic known as a veronica or vernicle. The flower is named after St. Veronica through a supposed resemblance of the flower's blossom to Christ's face. Of the flowers used in the church, the speedwell is dedicated to St. Veronica. The common name speedwell is an old equivalent of goodbye, a name suggested in 1884 by Richard Folkard because of the habit of the plant's corollas to drop off and fly away as soon as the flowers are collected.

The unusual blue color of speedwell caught the fancy of many writers. Ebenezer Elliott, a master metal founder in Sheffield, England, was also a writer of poems, many of which were collected in the *Corn Law Rhymes* (1830). In "Spirits and Men," an epic poem about the world before the biblical flood, he wrote, "Again a child, when childhood roved, I run/While groups of speedwells with their bright blue eyes,/Like happy children, cluster in the sun." Those same bright blue eyes inspired Wordsworth in "Flowers" from *The River Duddon* (1820) in which he referred to veronica by one of its Old English names: "The trembling eyebright showed her sapphire blue."

That Tennyson cherished the speedwell is apparent in "Spring," a part of the elegy *In Memoriam* (1850). He listed the flowers that he loved to summon the new life of spring:

> Bring orchis, bring the foxglove spire,
> The little speedwell's darling blue,
> Deep tulips dashed with fiery dew,
> Laburnums, dropping-wells of fire.

Ebenezer Elliott wrote in "The Winter Speedwell":

> Ye winter flowers, whose pensive dyes
> Wake when the summer's lily sleeps!
> Ye are like orphans, in whose eyes
> Their low-laid mother's beauty weeps.

In the language of flowers the speedwell stood for female fidelity. Victorian writer Charlotte Mary Yonge, who wrote frequently of siblings and large families, used this symbolism in *Countess Kate* (1862): "The young lady . . . delicately blue and white, like a speedwell flower."

In an essay titled "Wild Flowers" from *The Open Air* (1885), describing the wild flowers of his native Wiltshire, England, writer and naturalist Richard Jefferies told of trying to correctly identify the speed-well:

> Blue veronica was the next identified, sometimes called germander speedwell, sometimes bird's-eye, where leaves are so plain and petals so blue. Many names increase the trouble of identification, and confusion is made certain by the use of various systems of classification. The flower itself I knew, its name I could not be sure of—not even from the illustration, which was incorrectly coloured; the central white spot of the flower was reddish in the plate.

STOCK

Their scent and beauty joyn

The stock or stock-gilliflower is a member of the *Matthiola*, which consists of about fifty-five species of annuals, perennials, or subshrubs in the mustard family, the Cruciferae. Stock typically has large purple or white flowers that are often sweetly scented. Their native distribution is through western Europe, central Asia, and South Africa. Brompton stock, *M. incana* (formerly *M. annua*), includes 'Annua', a fast-growing selection called ten-weeks stock. *Matthiola longipetala* subsp. *bicornis* is the night-scented stock (formerly *M. bicornis*). The numerous hybrids and cultivars are generally treated as annual bedding plants: the ten weeks stock; the dwarf, bushy, Trisomic, seven-weeks races; and the perpetual, year-round-blooming dwarf form. The Virginia stock, sometimes called Malcolm stock, is *Malcolmia maritima*, a plant related but distinct enough to now be classified in a separate genus.

Matthiola is named after an Italian botanist and physician of the sixteenth century, Pierandrea Mattioli, who wrote and illustrated with woodcuts a knowledgeable work on the Greek physician Dioscorides. The species name *incana* means gray; *annua* refers to its annual growth habit; and *longipetala* means long-petaled. The allied genus *Malcolmia* is named after William and William Malcolm, father and son nurserymen in London in the late 1700s and the early 1800s. The species name *maritima* points to the preferred coastal habitat.

The common name stock suggests the woody stalk produced by some species. Since the time of its introduction to England in the early 1500s from the Mediterranean area, *Matthiola* has been called stock-gilliflower, meaning a gilliflower with a wooden shoot, to distinguish it from the clove-scented gilliflower of the genus *Dianthus*, which has a scent similar to *Matthiola*. The name gilliflower itself is a corruption of *giroflée*, the French name for the dianthus. The name stock-gilliflower was formalized in 1530 in Jehan Palsgrave's English-French grammar, *Lesclarcissement de la langue françoyse*, as "stocke gyllofer, *armorie bastarde*." Englishman William Turner, perhaps borrowing from Palsgrave, wrote of "Purple and blew stock-gelefloures" in *The Names of Herbes* (1548), a compilation of the names of plants in several languages. Through common usage over time, the full name stock-gilliflower became truncated to simply stock.

English philosopher and essayist Francis Bacon described natural phenomena in his essay on gardens *Sylva Sylvarum* (1626). He distinguished the stock-gilliflower from other plants: "In April follow the double white violet, the wall-flower, the stock-gilliflower, the cowslip, the fleur-de-luces, and lilies of all natures."

John Parkinson, herbalist to the king, described both double and single forms of stock in his *Paradisi in Sole, Paradisus Terrestris* (1629). He was not impressed with them, or had not studied them closely, since he wrote that they have little or no beauty and no good scent. He pointed out that the stock was also known incorrectly as *Leucojum* or *Viola alba*, but "we call it generally, Stockegilloflower (or as others doe, stockegillouer) to put a difference between them, and the Gilloflowers and Carnations, which are quite of another kindred, as shall be shewne in place convenient."

Seeing the plant as a symbol of everlasting beauty, the Victorian language of flowers took its meaning for stock not from Parkinson, but from the earlier English herbalist John Gerard, who wrote in his

Herbal or General History of Plants (1633 revision by Thomas Johnson) that the "stocke gillofloures" have double flowers "which are of divers colours, greatly esteemed for the beautie of their flourer, and pleasant sweet smell." René Rapin also wrote of these tender plants and their aromatic scents, giving directions for preserving them through the cold months:

> In other borders stocks their Heads divide,
> Whose curling Blooms are now expanded wide,
> Their Leaves with various Red diversify'd:
> But ah! Preserve 'em from too rude an Air,
> Their Scent and Beauty joyn to court your Care.
> And since they will not Winter's Cold endure,
> The tender Plants from threat'ing Winds secure,
> From Danger free they may in Pots be set,
> That of November mourn with drenching Wet
> You may within Doors lodge 'em safe from Harm,
> And keep in Stoves the tender Beauties warm.

SUNFLOWER

I lift my golden orb

The common sunflower of gardens is *Helianthus annuus*, with several well-known cultivated forms including 'Russian Giant', 'Flore Pleno', and 'Citrinus'. The natural geographic distribution of the sunflower is limited to the Americas, where there are seventy species of this herbaceous genus in the Compositae (Asteraceae). The genus consists of both annuals and perennials that bloom primarily in late summer to early fall. Some are valued for the oil extracted from their seed for cooking, and for the seed itself for birds and other wildlife. Other garden species include the golden-flowered *H. angustifolius*, called the swamp sunflower, that grows up to 2 meters (7 feet) tall, and *H. tuberosus*, the Jerusalem artichoke. The latter is not a true artichoke (*Cynara*, also a composite), but its tuberous roots are edible.

Greek mythology explains the first appearance of the sunflower. A water nymph named Clytie was enamored of the sun god Apollo, whom she had seen one day on a visit to Mt. Olympus. Every day she pined for him as she watched him move across the sky, but her affection was spurned and unreturned because Apollo was interested in Calliope, the muse of epic poetry. Despairing, Clytie took neither food nor drink. After nine days the gods took pity on her and changed her into a flower whose face would daily follow the course of the sun across the heavens. The sunflower in this myth could not have been the modern helianthus, because it is native only in North and South America and would not have been known to the ancients. The reference is probably to species of *Heliotropium* (in the Boraginaceae), a plant that is botanically unrelated to helianthus but which also exhibits strong heliotropism.

The origin of the botanical name *Helianthus* is from the Greek words *helios*, for sun, and *anthos*, for flower. The name may arise from the heliotropic behavior of the plant, which causes it to orient its flowers to the sun as the light moves across the sky. It is possible that Linnaeus, who cataloged specimens sent to him from Virginia, Peru, and Mexico, gave the Latin name to the sunflower because of the general resemblance of the flower head to the rays of the sun. Herbalist John Gerard called the sunflower the marigold of Peru:

> The floure of the Sun is called in Latine *Flos Solis*, taking that name from those that have reported it to turne with the Sun, the which I could never observe, although I have endeavoured to finde out the truth of it, but I rather think it was so called because it doth resemble the radiant beames of the Sun . . . . These plants do grow of themselves without setting or sowing, in Peru, and in diverse other provinces of America, from whence the seeds have beene brought into these parts of Europe.

Because of its radial arrangement of petals and its yellow color, the sunflower was held in high esteem by Incan priests as a symbol of the nobility of the sun god. At the Temple of the Sun, honored rulers and sacrificial virgins were crowned with representations of sunflowers made of pure gold. The Incas reproduced the sunflower in various art forms, such as weaving, pottery, and jewelry-making. Sunflower jewelry was worn by the priests and virgins of the sun god. The celebrated Island of the Sun in Lake Titicaca on the border between Bolivia and Peru is the mythological site where the first man and woman of Incan legend came to earth. Celebrations of the sun and the sunflower are believed to have been held there.

The Spanish conquistador Hernando Cortés carried the helianthus from Central America to Europe in the 1500s, leading to its use in European religious symbolism. The Old English church recommended decorating with the sunflower on St. Bartholomew's day, 24 August, because it represented constancy and devotion, symbolism that was retained in the Victorian language of flowers. A rhyming calendar of the fifteenth century explained, "And yet anon the full sunflower flew,/And became a starre for Bartholomew." An anonymous ode highlighted the devotion of the sunflower:

> Emblem of constancy, whilst he is beaming,
> For whom is thy passion so steadfast, so true;
> May we, who of faith and of love are aye dreaming,
> Be taught to remember this lesson by you.

In literature the "sunflower" appeared in Chinese works as early as the eighth century. In ancient Chinese cultures, the sunflower was a symbol of longevity and its seeds were the food of immortality. Those references, however, cannot be applied to the American native helianthus. Probably a European or Asian composite was meant, perhaps a yellow aster or chrysanthemum. Chinese poet Wang Wei mentioned a sunflower in "My Lodge at Wang-Ch'uan After a Long Rain," translated by Witter Bynn:

> The woods have stored the rain, and slow comes the smoke
> As rice is cooked on faggots and carried to the fields;
> Over the quiet marsh-land flies a white egret,
> And mango-birds are singing in the full summer trees . . .
> I have learned to watch in peace the mountain morning
>   glories,
> To eat split dewy sunflower-seed under a bough of pine,
> To yield the post of honour to any boor at all . . .
> Why should I frighten sea-gulls, even with a thought?

As the sunflower made its way to European gardens, it also began appearing in poetry. In "Ah! Sunflower," William Blake saw the plant as a kind of clock:

> Ah! Sunflower! weary of time,
> Who countest the steps of the sun;
> Seeking after that sweet golden clime,
> Where the traveller's journey is done.

Percy Bysshe Shelley, in his "Sunflower," implored the bloom not to be so constant:

> Light-enchanted sunflower! Thou
> Who gazest ever true and tender
> On the sun's revolving splendor,
> Follow not his faithless glance
> With thy faded countenance.

In a brooding poem titled "Song" (1830), Alfred, Lord Tennyson, described the last hours of the day, when evening replaces the light of the sun:

> Heavily hangs the broad sunflower
> Over its grave i' the earth so chilly;
> Heavily hangs the hollyhock,
> Heavily hangs the tiger-lily.

Thomas Moore recalled the story of Clytie's constancy and unrequited love for Apollo:

> The heart that has truly loved never forgets,
> But as truly loves on to the close;
> As the Sunflower turns on her god when he sets,
> The same look that she turned when he rose.

In *A Tale of Two Cities* (1859), Charles Dickens used the sunflower to describe the lawyer Mr. Stryver:

A favourite at the Old Bailey, and eke at the Sessions, Mr.
Stryver had begun cautiously to hew away the lower staves
of the ladder on which he mounted. Sessions and Old Bailey had now to summon their favourite, specially, to their
longing arms; and shouldering itself towards the visage of
the Lord Chief Justice in the Court of King's Bench, the
florid countenance of Mr. Stryver might be daily seen,
bursting out of the bed of wigs, like a great sunflower
pushing its way at the sun from among a rank garden-full
of flaring companions.

In "Rudel to the Lady of Tripoli" by Robert Browning, a twelfth-
century troubadour from Provence named Geoffrey de Rudel spent
all his efforts searching for the Countess of Tripoli, whom he had
never seen. Lying on his deathbed, wearing the emblematic sunflower
typical of troubadours, he speaks of the sunflower and of never having seen the face of his love:

> With all a flower's true graces, for the grace
> Of being but a foolish mimic sun,
> With ray-like florets round a disk-like face—
>
> . . . .
>
> That I French Rudel chose for my device
> A sunflower outspread like a sacrifice
> Before its idol.

Erasmus Darwin was a Cambridge-educated physician and naturalist who maintained a physic garden. He illustrated the botanical
system of Linnaean classification in a book-length poem called *The
Loves of the Plants* (1789), which he wrote in heroic couplets, imitative
of Alexander Pope. He told of the goddess Flora who descended to
earth to explain various natural phenomena, including the "harems"
of stamens and pistils used by Linnaeus. Included is a forerunner of

evolutionary theory, which years later inspired the writer's grandson, Charles Darwin. Of the "great helianthus" he wrote:

> With zealous step he climbs the upland lawn,
> And bows in homage to the rising dawn;
> Imbibes with eagle eye to the rising ray,
> And watches, as it moves, the orb of day.

The early Victorian poet Leticia Elizabeth Landon saw the hopeless and unhappy side of the sunflower and its myths. She wrote in "The Sunflower,"

> Symbol of unhappy love
> Sacred to the slighted Clytie:
> See how it turns its bosom to the sun;
> And when the dark clouds have shadowed it,
> Or night is on the sky, mark how it folds its leaves
> And droops its head, and weeps sweet tears of dew:
> The constant flower!

The education of young girls, the suffrage movement, antivivisection, and the education of the mentally retarded were nineteenth-century causes championed in England by poet and essayist Dora Greenwell. Christina Rossetti was "quite struck with [Greenwell's] large-mindedness" on various social issues. Greenwell's version of "The Sun-flower" (1848) used the flower's compulsive daily movements as a metaphor for sexual desire and slavery, probably a highly controversial message for the era:

> Till the slow daylight pale,
> A willing slave, fast bound to one above,
> I wait; he seems to speed, and change, and fail;
> I know he will not move.
>
> I lift my golden orb
> To his, unsmitten when the roses die,

And in my broad and burning disk absorb
The splendours of his eye.
His eye is like a clear
Keen flame that searches through me: I must droop
Upon my stalk, I cannot touch his sphere;
To mine he cannot stoop.

I win not my desire,
And yet I fail not my guerdon; lo!
A thousand flickering darts and tongues of fire
Around me spread and glow.

All ray'd and crown'd, I miss
No queenly state until the summer wane,
The hours flit by; none knoweth of my bliss,
And none has guessed my pain.

I follow one above,
I track the shadow of his steps, I grow
Most like to him I love
Of all that shines below.

The word sunflower was also applied figuratively to a person of re-
splendent beauty. Lord Byron wrote of "Neuha, the sun-flower of the
island daughters" in *The Island* (1823), a poem about the mutiny of the
HMS *Bounty*. It tells of earthly paradise and idyllic love between Neuha
and Torquil, newlyweds on the tropical isle of Toobonai before "the
white man landed!—need the rest be told?" William Makepeace
Thackeray wrote in *Vanity Fair* (1848), "There are garden ornaments as
big as brass-warming pans, that are fit to stare the sun itself out of
countenance. Miss Sedley was not of the sunflower sort."

Vincent van Gogh considered the sunflower a symbol of light, renewal, and health; he painted sunflowers and decorated his house and
bedrooms with them. In the late nineteenth century as van Gogh
painted a series of sunflowers in The Yellow House at Arles in Pro-

vence (where he cut off a part of his own ear during a fit of depression), the sunflower (and the lily) became a symbol of the Aesthetic Movement, or the belief in "art for art's sake," that was championed by Théophile Gautier, Charles Baudelaire, Gustave Flaubert, and Oscar Wilde. The movement proclaimed that art should not serve any purpose beyond itself, creating, perhaps inadvertently, the notion that the artist is an especially sensitive person of superior intellect.

In the last decade of the twentieth century, the sunflower is enjoying a sort of helianthus resurrection as van Gogh's images and those of others are widely imprinted on *objects d'art* and utilitarian items. Oscar Wilde may have had such gaudy kitsch in mind when he wrote "Le Jardin" in "Impressions (I.)":

> The gaudy leonine sunflower
> Hangs black and barren on its stalk,
> And down the windy garden walk
> The dead leaves scatter,—hour by hour.

TULIP

The varied colors run

The 100 species of this bulbous perennial of the Liliaceae are native to the temperate regions of central Asia. Tulips are primarily spring flowering, usually solitary flowers on stems as short as 8 centimeters (3 inches) or as tall as 1 meter (3 feet). All tulips belong to the genus *Tulipa*, which currently encompasses fifteen subdivisions of garden tulips and more than 2300 hybrids and cultivars listed on the international tulip registry. Well-known species are *T. kaufmanniana*, *T. fosteriana*, and *T. greigii*. The tall Darwin hybrids result from a cross between *T. fosteriana* and other hybrids. The prized Rembrandt division tulips, frequently depicted in Dutch paintings, have striped or feathered markings on its petals due to a viral infection. The lady tulip, *T. clusiana*, and *T. humilis* are also commonly grown.

In its native Turkey the tulip was known as *lalé*, but foreign travelers noticed the flower's resemblance to a turban. Thus, the common name tulip is the Latin translation of the Turkish word for turban, *tulbend*. As John Gerard explained in *The Herbal or General History of Plants* (1633 revision by Thomas Johnson), "Tulipa, or the Dalmatian cap, is a strang and forraine flower . . . After [the Tulipa] hath beene some fewe daies floured, the points and brims of the flower turne backward, like a Dalmatian or Turkes cap, called Tulipan, Tolepan, Turban, and Turfan, whereof it tooke his name."

In the language of flowers, the red tulip suggested a

declaration of love, and the yellow tulip declared hopeless love. The variegated or streaked tulip said, "You have beautiful eyes." In the church, tulips were appropriate for use on 1 May, the day dedicated to St. Philip and St. James.

Tulips were grown in Turkey and included on decorative motifs before they were introduced into Europe in the mid-1500s by Ogier Ghislain de Busbecq, the Austrian ambassador to Suleiman the Magnificent of the Turkish Empire. Carolus Clusius, a French botanist who maintained the imperial gardens in Vienna and after whom the species *Tulipa clusiana* is named, obtained tulip seed from de Busbecq, germinated them, and sent bulbs to England in 1578. After Clusius became a botany professor at the University of Leyden, he took bulbs with him to Holland in 1593 and planted them at the university gardens.

Excitement—called tulipomania—spread rapidly throughout Europe over the distinctive imported flower, focusing especially in Holland. It reached its extravagant height in the 1630s with speculators investing in tulip bulbs hoping to realize huge profits from their growth and multiplication. The price of single bulbs skyrocketed—some sold for more Dutch florins than the average citizen earned in a lifetime. The financial frenzy led to ultimate crash and ruin for most in 1637, when the government began regulating the tulip market. The Dutch continue to lead the world in growing and exporting tulips. Their love of the flower was evidenced in its inclusion in the great art, particularly the paintings, of the 1600s. Tulip motifs were carried across the ocean to North America, where they appeared prominently in the art of German immigrants who settled in Pennsylvania, the so-called Pennsylvania Dutch.

Praise for this turbaned flower has found its way into poetry, song, and literature. English folktales tell of pixies using tulips as cradles for their babies. One night a woman found the sleeping babes and delightedly planted more tulips. The tulips flourished, and in time there were enough cradles for all the wee folk. When the woman died, a miserly farmer plowed up the tulips and planted parsley. The angry pixies stole out at night and nipped the roots of the parsley. Nothing thrived in the garden again, but tulips grew plentifully on the grave of the old woman.

The beauty of the tulip inspired many anonymous poets to extol its virtues. For one, it surpassed all competition:

> Vainly in gaudy colours drest,
> 'Tis rather gazed on than caressed.
> For brilliant tints to charm the eye,
> What flow'r can with the Tulip vie?

The Jesuit poet René Rapin speculated that flowers were really nymphs that the gods had metamorphosed. The tulip he thought was a Dalmatian nymph changed as she fled the advances of Vertumnus:

> Six gaudy leaves a painted cup compose,
> On which kind nature every dye bestows;
> For though the nymph's transformed, the love she bore
> To colors still delights her as before.
> The Tulip which with gaudy Colours stained,
> The Name of Beauty to her Race has gain'd;
> For whether she in Scarlet may delight,
> Chequer'd and streak'd with Lines of glittering white,
> Or ting'd all o'er with purple, charms our Sight;
> Or widow-like beneath a sable Veil,
> Her purest Lawn may artfully conceal,
> Or emulates the vary'd Agate's veins,
> She beauties prize from ev'ry flow'r obtains.

Robert Herrick lamented the passing of young women's virginity by comparing them to the flowers in "To a Bed of Tulips":

> Bright tulips, we do know
> You had your coming hither;
> And fading-time does show,

That ye must quickly wither.

. . . .

Come Virgins, then and see
Your frailties; and bemone ye;
For lost like these, 'twill be,
As Time had never known ye.

James Thomson of Scotland saluted the tulip's brilliant colors in
"Spring" (1728) from *The Seasons*:

Then comes the Tulip race, where Beauty plays
Her idle freaks: from family diffused
To family, as flies the father dust,
The varied colours run.

And Leticia Elizabeth Landon gave a tribute to tulips in "A History
of the Lyre," in which the beautiful Roman lute player Eulalie, orphaned, lonely, and sad, thinks of flowers that bring sweetness and
beauty:

Again I'll borrow Summer's eloquence.
Yon Eastern tulip—that is emblem mine;
Ay! it has radiant colours—every leaf
Is a gem from its own country's mines.

Though most impressions of the tulip are more than favorable,
Rector Thomas Fuller, who had a fondness for writing in aphorisms,
saw them differently. In his *Speech of Flowers* (ca. 1660) he gave voice to
a jealous rose who, sneering bitterly, complains of the tulip:

There is lately a flowre—shal I call it so?—in courtesie I
will tearme it so, though it deserve not the appellation, a
Toolip, which hath engrafted the love and affections of
most people unto it; and what is this Toolip? A well complexion'd stink, an ill favour wrapt in pleasant colours; as

for the use thereof in physick, no physitian hath honoured
it yet with the mention, nor with a Greek, or Latin name, so
inconsiderable hath it hitherto been accompted; and yet this
is that which filleth all gardens, hundred of pounds being
given for the root thereof, whilst I the Rose, am neglected
and contemned, and conceived beneath the honour of noble
hands, and fit only to grow in the gardens of yeomen.

VERBENA OR VERVAIN

A wreath of vervain heralds wear

The wild or common vervain of southern Europe is *Verbena officinalis*, formerly highly valued for its medicinal properties. Most of the other more than 250 species of *Verbena* in the Verbenaceae are natives of tropical and temperate regions of North and South America. The popular South American *V. bonariensis* may reach a height of 2 meters (7 feet), its thin wiry stems topped by small clusters of light purple flowers. *Verbena ×hybrida* consists of a large number of hybrid garden verbenas resulting from crosses between several species, including *V. incisa*, *V. peruviana*, *V. phlogiflora*, and *V. teucroides*. This hybrid complex includes such cultivars as 'Homestead Purple', 'Silver Anne', and 'Sissinghurst'.

Verbena was the classical Roman name for any plant used medicinally or as an altar decoration. The term collectively referred to laurel, myrtle, or olive and is assumed to have included *V. officinalis*, the apothecary verbena. *Verbena officinalis* is the only species that could have been known by the Romans, for all the other verbenas were growing only in the as-yet-undiscovered New World. The species *V. bonariensis* was the first to reach Europe from South America in the early 1700s; its name is Latin for Buenos Aires, the source of the seed. One source suggests that the word verbena is a corruption of *herbena*, the truncated Latin *herb bona*, meaning the good plant. The English name vervain comes from the thirteenth-century Old French words *verveine* and *vervaine* for "green bough."

Vervain was called the sacred herb by the Greeks, who venerated it and ascribed numerous wondrous properties to it, including the ability to mollify enemies and help reconcile differences. Roman ambassadors used it in peace treaties; a herald called *verbenarius* would carry boughs of verbena to the negotiations. Vervain was valued by the Druids, second only to mistletoe. They used it in divinations and all manner of mystic ceremonies. They offered sacrifices before cutting the plant and afterwards poured honey on the ground in atonement for taking the precious herb from the earth. Druids were allowed to collect vervain only in the springtime before the rising of the Great Dog Star, when neither the sun nor the moon would be above the horizon to witness the taking. Some sources state that the plant should be collected only at the full moon.

Verbena officinalis was also called Juno's tears. As the stories go, Juno, or Hera in Greek mythology, angry and jealous because Jupiter, or Zeus, was paying too much attention to Callisto, turned the beautiful Callisto into a bear. While wandering in the forest Callisto's son saw the bear and killed it with a spear, not knowing that it was his own mother. Jupiter placed the bear and her son in the night sky as the Great Bear and the Little Bear. Juno was only further enraged and hastened to the sea to seek help and advice from Oceanus, along the way dripping tears that turned to vervain plants. But old Oceanus had no interest in Juno, and the night sky continues to honor Callisto and her son.

Nicholas Culpeper, apothecary and herbalist, ascribed numerous attributes to the common vervain and called it the herb of Venus. It was most effective for ailments of the womb but was used to treat a plethora of other problems, such as jaundice, dropsy, gout, worms, and hemorrhoids. Because of its wide use as a simple, an herb used for medicinal purposes, vervain was known as simpler's joy. Among native North Americans, *Verbena canadensis* was known to the Pawnee as pleasant-dream plant and to the Omaha as stomachache plant. Sir William D'avenant, writing in the seventeenth century, alluded to vervain's curative powers in "Gondibert" (1650):

> Black melancholy rusts, that fed despair
> Through wounds' long rage, with sprinkled vervain cleared;
> Strewed leaves of willow to refresh the air,
> And with rich fumes his sullen senses cheered.

One of vervain's apothecary uses was to "cool the blood." In *Tristram Shandy*, a novel written between 1759 and 1767 by Laurence Sterne, Uncle Toby was advised to moderate his passions by taking an infusion of vervain prior to sitting in close quarters with the Widow Wadman.

In the pastoral tragi-comedy *The Faithful Shepherdess* (1609), collaboratively written by Englishmen Sir Francis Beaumont and John Fletcher, Clorin, a shepherdess, pines over her buried love, "the truest man that ever fed his flockes," as she sorts herbs and describes their healing benefits. She ends her list with the vervain:

> And thou, light vervain, too, thou must go after,
> Provoking easy souls to mirth and laughter;
> No more shall I dip thee in water now
> And sprinkle every post, and every bough
> With thy well-pleasing juice, to make the grooms
> Swell with high mirth, and with joy all the rooms.

Verbena had an additional presence in religious legends. The Rev. John White wrote in 1624: "Hallowed be thou, vervain, as thou growest in the ground,/For on the Mount of Calvary thou first wert found." Vervain was also said to have staunched the wounds of the crucified Jesus.

Verbena figured in the descriptions of ancient ceremonies. Dryden translated from Virgil's *Pastoral*, "Bind those Altars round With Fillets; and with Vervain strow the Ground." Michael Drayton wrote in *The Muses Elyzium*:

> A wreath of vervain heralds wear
> Amongst our garlands named,
> Being sent that dreadful news to bear
> Offensive war proclaimed.

And Ben Jonson mentioned the vervain in *Sejanus his Fall* (1603): "Bestow your garlands: and, with reverence, place the vervin on the altar."

Often writers have been struck by the wonderful scent that the verbena exudes, as was Elizabeth Barrett Browning writing in *Aurora Leigh* (1856): "Like Sweet verbena which, being brushed against,/Will hold us three hours after by the smell,/In spite of long walks on the windy hills."

Frederick Tennyson, elder brother of Alfred, mentioned in "Psyche," his poem of the Greek isles that recalls the rural features of the Lincolnshire wolds: "Garden sweets,/Jasmin and vervains, and old lavender/And savorous herbs."

In *Le Vicomte de Bragelonne* (1848–1850), a six-volume historical romance that is the sequel to *The Three Musketeers* by Alexandre Dumas *père*, the presence of Mademoiselle Louise De La Vallière, maid of honor to the Duchesse d'Orléans and mistress to King Louis XIV, is betrayed by the wearing of La Vallière's favorite vervain scent. She wears it because it first led her to a secret staircase and a room where she met the king.

Vervain and verbena are endowed with a vast number of benefits. Remember them for curing scrofula, arresting the diffusion of poison, neutralizing rabid mammals, and more. Carry them into arenas requiring international diplomacy, and utilize their wide range of sentiments from the language of flowers: reconciliation, sensibility, purity, and guiltlessness.

VIOLET, PANSY & VIOLA

May boast itself the fairest flower

About 500 species and countless cultivars and hybrids of the well-known genus *Viola* all have five petals, five sepals, and five stamens. These members of the Violaceae are widely distributed in northern temperate regions, but some species are native to New Zealand, the Amazon, and the Andes. Popular in the garden are violas and pansies. Pansies are hybrids known as *Viola ×wittrockiana*, derived from *V. tricolor* and other species. In addition a large number of cultivated varieties and hybrids are derived from crosses of *V. lutea*, *V. amoena*, and *V. cornuta*. Among the most popular species are *V. tricolor*, with numerous common names including heartsease and Johnny-jump-up; *V. odorata*, sweet violet; and *V. cornuta*, viola. Among the wild violets of the United States are the bird's foot, *Viola pedata*, with deeply divided leaves, the woolly blue violet, *V. sororaria* (formerly *V. papilionacea*), and the western dog violet of the Rocky Mountains, *V. adunca*.

The modern name violet derives from the early Latin name *viola*, which appears to be related to *io* or *ion*, the plant's ancient Greek name. According to mythology, Zeus seduced the priestess Io, making his wife Hera furious. In order to escape her fury, Io was transformed into a white heifer, and Zeus caused sweet violets to spring from the ground for her to feed upon. He named the flowers ion. An earlier version of the Greek spelling for ion contained an initial letter, the digamma, similar to a *v* or *w*, forming *vion*. Hence, both the English and Latin names derive from it. In Greek the name also generally applied to any sweet-smelling flowers. Now, the plant and the color have become synonymous, although not all violets are violet colored. The word iodine, the violet-colored tincture, has a related etymological origin.

The pansy, also a member of the genus *Viola*, is named from the French *pensée*, which means thought. Appropriately, in the language of flowers it symbolized remembrance. Shakespeare used this meaning in act 4 of *Hamlet* (1601) when Ophelia says, "There's pansies, that's for thoughts." Later she speaks of violets again: "There's a daisy: I would give you some violets, but they all withered when my father died."

The violet is among the most beloved of all plants. It was known in classical times and is frequently mentioned by Homer and Virgil. In the Middle Ages the prize awarded to the best poet was a golden violet. The violet was Queen Victoria's favorite flower, and Shakespeare alluded to it numerous times.

Shakespeare called *Viola tricolor* love-in-idleness in *A Midsummer Night's Dream* (1600), but *V. tricolor* and its relatives go by many appellations. An ancient Latin name was *Flos Jovis*, or Jove's flower. An old Italian name is *fammola*, or "a little flame." The English have a score of other names, including jump-up-and-kiss-me-quick, herb trinity, and cuddle-me-to-you.

According to an old proverb, when roses and violets flourish in the autumn and winter, it is a sign of plague or pestilence for the coming year. Interpreted in practical terms, the proverb explains that a mild winter that allows spring flowers to bloom early is less healthy than a cold winter that kills bacteria and viruses.

When Napoleon Bonaparte, Emperor of France, abdicated under pressure from the allied forces led by the first duke of Wellington, he was banished to the island of Elba in 1814. He told his followers that he would return when the violets were again in season. As a result of this statement, violets became a secret symbol by which his supporters recognized each other. To test new acquaintances, one asked *"Aimez-vous les violettes?"* (Do you like violets?). If the answer were a noncommittal *"Eh, bien"* (Well . . .), a confederate conspirator would respond by completing the phrase with *"Elles reparaîtront au printemps"* (They will reappear again in the spring). If the response were only *"Oui"* (Yes), it was inferred that the person was not a Napoleonic adherent and the conversation would proceed cautiously. Napoleon's underground supporters often wore violet-colored items, such as jewels or watch ribbons, to identify themselves to each other. English poet Lord Byron gave voice to Napoleon on his banishment to Elba in "Napoleon's Farewell":

> Farewell to thee, France! But when Liberty rallies
> Once more in thy regions, remember me then—
> The Violet still grows in the depths of thy valleys;
> Though withered, thy tears will unfold it again.

When Napoleon escaped from Elba after ten months and appeared at the Tuilleries on 20 March 1815, he was presented with a bouquet of violets. The flower became so symbolic of him that after his great defeat at Waterloo on 18 June 1815 and after the restoration of Louis XVIII to the throne, the violet became a symbol of sedition and was dangerous to wear even in a buttonhole. The violet became Napoleon's emblem and the source of his noms de plume Père la Violette and Corporal Violette.

The violet came to mean faithfulness in the Victorian language of flowers. William Hunnis's poem from the end of the 1500s attests to the strong roots of that symbolism. His poem titled "In a Nosegay always sweet, for lovers to send for tokens of love at New Year's Tide, or for fairings" says,

> Violet is for faithfulness

Which in me shall abide;
Hoping likewise that from your heart
You will not let it slide.
And will continue in the same
As you have now begun,
And then for ever to abide,
Then you my heart have won.

Robert Herrick explained the naming of heartsease (*Viola tricolor*) in "How Pansies or Hearts-ease Came First":

Frolic virgins, once these were,
Over loving, living here,
Being here their ends denied,
Ran for Sweethearts mad and died.
Love, in pity of their tears,
And their loss in blooming years,
For their restless here spent hours
Gave them Heart's-ease, turn'd to Flowers.

In Shakespeare's *A Midsummer Night's Dream*, Oberon, king of the fairies, enlists Puck's help in a scheme to ridicule his queen, Titania, by making her fall in love with Bottom, a traveling artisan:

It fell upon a little western flower,
Before milk-white, now purple with love's wound,
And maidens call it, love-in-idleness.
Fetch me that flower; the herb I showed thee once:
The juice of it on sleeping eyelids laid
Will make or man or woman madly dote
Upon the next live creature that it sees.
Fetch me this herb: and be thou here again.
Ere the leviathan can swim a league.

In "On a Faded Violet" (1818), Percy Bysshe Shelley lamented,

> The odour from the flower is gone
> Which like thy kisses breathed on me;
> The colour from the flower is flown
> Which glowed of thee and only thee!

In "Heart's-Ease," Mary Botham Howitt considered the name and history of the flower:

> Heart's-Ease! one could look for half a day
> Upon this flower, and shape in fancy out
> Full twenty different tales of love and sorrow,
> That gave this gentle name.

Numerous paeans and odes are written specifically to the beauty of the violet, which is often compared by the poets to their loved ones' sweetest qualities. Robert Herrick in his "To Violets" called violets the "maids of honour" that bloom early in the spring; the poem closes,

> You're the maiden posies,
> And so graced
> To be placed
> 'Fore damask roses.

Sir Walter Scott noted in "The Violet (The Fairest Flower)" (1797),

> The violet in her greenwood bower,
> Where birchen boughs with hazels mingle,
> May boast itself the fairest flower
> In glen, or copse, or forest dingle.

Walter Savage Landor wrote,

> One Pansy, one, she bore beneath her breast,
> A broad white ribbon held that Pansy tight.
> She waved about nor looked upon the rest,
> Costly and rare; on this she bent her sight.

In "The Yellow Violet," William Cullen Bryant ended the poem with

> And when again the genial hour
> Awakes the painted tribes of light,
> I'll not o'erlook the modest flower
> That made the woods of April bright.

But Brian Waller Proctor, under the pseudonym Barry Cornwall, was perhaps warmest in praise of the violet:

> I Love all things the seasons bring,
> All buds that start, all birds that sing,
> All leaves, from white to jet,
>
> . . . .
>
> But chief the violet.

William Wordsworth wrote of "Lucy" (1798) who "dwelt among the untrodden ways," comparing her to a violet:

> A violet by a mossy stone
> Half hidden from the eye!
> —Fair as a star, when only one
> Is shining in the sky.

Wordsworth used the flower again, this time as the pansy, in his "Ode: Intimations of Immortality from Recollections of Early Childhood" (1802). In this profound work Wordsworth revealed a belief in pre-existence. His childhood memories seemed to hint at the great knowledge and understanding that he once had but could no longer grasp:

> The Pansy at my feet
> Doth the same tale repeat:
> Whither is fled the visionary gleam?
> Where is it now, the glory and the dream?

Aubrey De Vere grew nostalgic over the "Lone flower of many names, the wind sweeps o'er thee" in "To a Wild Pansy," but Mrs. C. W. Earle's love for the sweet violet is clear from her diary entry for 14 March 1899, published in *More Pot-Pourri from a Surrey Garden*:

My garden is now full of the old wild sweet Violet (*Violet odorata*) of our youth before even the 'Czars' came in, much less the giant new kinds. I have an immense affection for this Violet, with its beautiful intense colour and its delicate perfume. It grew all about the Hertfordshire garden under the hedges, and little seedlings started up in the gravel paths, looking bold and defiant; but all the same they were rooted out by the gardener when summer tidying began. At the end of March or early April, when the rain comes, I divide up and plant little bits of these Violets everywhere, and they grow and flourish and increase under Gooseberry bushes and Currant bushes, along the palings covered with Blackberries, under shrubs—anywhere, in fact—and there they remain hidden and shaded and undisturbed all the summer. Where seedlings appear they are let alone all the summer and autumn till after flowering-time in spring. They look lovely, and brave these cold, dry, March days; but their stalks are rather short here, for want of moisture. If anyone wants to see this Violet to perfection, let him be in Rome early in March, as I once was, and let him go to the old English cemetery, where Keats lies buried and the heart of Shelley, and he will see a never-to-be-forgotten sight—the whole ground blue with the Violets, tall and strong above their leaves, the air one sweet perfume, and the sound (soft and yet distinct) of the murmur of spring bees.

WALLFLOWER

Flower in the crannied wall

The wallflower, or *Erysimum*, is a member of the Cruciferae, the mustard family. The eighty or more annual or perennial species of this genus are distributed in the Northern Hemisphere. *Erysimum cheiri* (formerly *Cheiranthus cheiri*) has the yellow flowers typical of the family, but *E.* 'Bowles' Mauve' bears rich mauve flowers in late winter. The Siberian wallflower, *E.* ×*allionii*, offers bright orange flowers.

The name *Erysimum* is from the Greek *eryo*, meaning "to draw out," a reference to the skin-irritating, even blistering aspects of some species, used medicinally to alleviate more serious inflammations. Ancient Greeks also called the plant erysimum. The former genus name *Cheiranthus* is from the Greek words *cheir*, for hand, and *anthos*, for flower, suggesting the plant's common use in bouquets and nosegays.

The precise botanical identity of the wallflower mentioned in literature is a matter of conjecture. Its name suggests flowers growing on walls, castles, and abandoned battlements. It is often also referred to as a gilliflower (in the sense of July flower, the name's etymological meaning) or simply "the flower in the crannied wall," as Tennyson called it. Robert Herrick associated romance with the "flower of the wall" growing on a castle in Scotland.

In the language of flowers, the wallflower suggested fidelity in adversity. This symbolism seems to stem from the time when troubadours roamed the countryside singing of misfortunes and lost loves, often wearing lapel flowers plucked from a passing wall. Herrick in "How the Wall-flower Came First, and Why So Called" gave an account of the origin of the wallflower and its use as a symbol for fidelity. He told of a young maiden who attempts to escape from the castle near the river Tweed in which she is imprisoned, aiming to join her lover and his attendant knights. Her misstep and untimely fall concludes the poem. The maiden dies upon the rocks, her frock transformed into green leaves and her bright bonnet turned into yellow blossoms stained with her blood:

> Up she got upon a wall,
> Tempting down to slide withal!
> But the silken twist untied,
> So she fell, and bruised, she died.
> Jove in pity to the deed,
> And her loving luckless speed,
> Turn'd her to this plant we call
> Now, the Flower of the Wall!

This metamorphosis became the basis for the wallflower's use as an emblem of fidelity and affection during adversity. In more traditional versions of the story, the lover became a troubadour and wandered the countryside lamenting his loss and wearing a yellow wallflower on his lapel or in his cap.

Many laudatory verses have been penned to this favored flower. Alfred, Lord Tennyson, saw the roots of knowledge in the flower:

> Flower in the crannied wall,
> I pluck you out of the crannies,
> I hold you here, root and all, in my hand,
> Little flower—but if I could understand
> What you are, root and all, and all in all,
> I should know what God and man is.

"The Wallflower," by an anonymous poet, finds both earthiness and spirituality in the flower:

> Cheerful 'midst desolation's sadness—thou—
> Fair flower, art wont to grace the moldering pile,
> And brightly bloom o'er ruin, like a smile
> Reposing calm on age's furrowed brow—
> Sweet monitor! an emblem sure I see
> Of virtue, and of virtue's power, in thee.
> For though thou cheerest the dull ruin's gloom,
> Still when thou'rt found upon the gay parterre,
> There thou art sweetest—fairest of the fair;—
> So virtue, while it robs of dread the tomb,
> Shines in the crown that youth and beauty wear,
> Being best of all the gems that glitter there.

And David Macbeth Moir, who often wrote under the pen name Delta, sometimes even signing with the symbol Δ, in "Wallflower" went the furthest with his praise:

> Sweet wallflower, sweet wallflower!
> Thou conjurest up to me
> Full many a soft and sunny hour
> Of boyhood's thoughtless glee;
>
> . . . .
>
> Of flowers first, last, and best!
> There may be grander in the bower,
> And statelier on the tree,
> But wallflower, loved wallflower,
> Thou art the flower for me!

WATERLILY
Her chalice reared of silver white

At least three genera of plants are referred to as waterlilies, the common name for various aquatic plants with large flowers. They are *Nymphaea*, *Nuphar*, and *Nelumbo*. *Nymphaea* and *Nuphar* are members of the Nymphaeaceae, and *Nelumbo* is a member of the Nelumbonaceae. Waterlilies may be hardy enough for temperate water gardens, or they may be so tender as to require near-tropical conditions for vigorous growth and flowering.

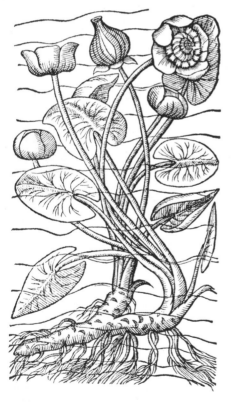

Nymphaea alba is the common white European waterlily, and *N. tuberosa* and *N. odorata* are from the eastern United States. Hardy and tropical cultivars and hybrids abound within the fifty species of this genus. The genus name is from the Greek *nymphe*, which means water nymph.

Nuphar lutea is the yellow waterlily or spatterdock, an aquatic plant with large floating leaves, of wide distribution in the Northern Hemisphere. The name of this genus is from *naufar*, the Arabic word for the plant.

Nelumbo, the water lotus, comprises only two perennial species, *N. lutea* and *N. nucifera*. The latter is the sacred lotus of the Far East. The genus contains at least a score of recognized cultivars, some with large fragrant flowers. *Nelumbo* should not be confused with the terrestrial plant *Lotus* of the Leguminosae, which bears small pea-like flowers. *Nelumbo* is the Sinhalese name for the plant.

The word lotus can refer to several different plants. As a waterlily, the lotus (*Nelumbo*) has been a sacred object for at least 5000 years. According to the writings of Mohammed, a lotus tree stands on the right hand of God in the seventh heaven. The ancient Egyptians dedicated the lotus to the god of the sun and adorned architectural columns with a lotus-like motif. In India the lotus was the symbol of Buddha. And Hindu images of Brahma show him sitting on a lotus, symbolizing his birth from out of the lotus-shaped navel of Vishnu. In yoga exercise, a body in the lotus position is said to resemble the shape of the lotus blossom. The Tibetan prayer chant "*Om mani padme hum*" translates as "I salute the jewel in the lotus."

Western cultures also draw upon the lotus as a symbol of purity and forgetfulness. Homeric legend tells of the lotus eaters who ate the fruit of the lotus tree. It caused them to forget their families and friends and to desire only to live their lives in idleness and luxury in Lotus-land. Tennyson wrote of those who succumbed to the plant in *The Lotos-Eaters* (1832):

> The Lotos blooms below the barren peak
> The Lotos blows by every winding creek
>
> . . . .
>
> In the hollow Lotos-land to live and lie reclined
> On the hills like Gods together, careless of mankind.

In the following myth, the blue lotus waterlily, *Nymphaea caerulea*, has a role in the genesis of the lotus shrub. A water nymph, Lotis, was smitten by the sight of the virile Heracles, who did not return her gazes. She pined too deeply for his love and died. Hebe, the goddess of youth and spring, turned Lotis into a waterlily with purple flowers. Later, Dryope and Iole, sister nymphs, along with Dryope's son, were picking flowers when they came upon the waterlily. When Dryope picked it, the stem dripped blood. Immediately her hands and feet began sprouting leaves and roots and her body became covered with rough bark. Dryope was turned into a lotus shrub and ever after bore the fruit of forgetfulness that was sought by the lotus eaters.

Waterlilies figure in several Native American legends including those of the Dakota, Iroquois, Pawnee, and Ojibway. In a Dakota story, Chief Red Strawberry Man was visited in a dream by Star-Maiden from the night sky. She told him that she was tired and wanted to come to earth to live among the Dakota. Red Strawberry Man decided to seek counsel from the tribe's advisor who lived across the lake, so he sent his son to get the wise one. The son paddled quickly in the darkness while the impatient maiden sat in the canoe with him. In his haste, the boy hit a log and the Star-Maiden tumbled into the water and disappeared. The next morning, growing at the spot where the maiden's light was extinguished was a waterlily with yellow flowers gleaming in the sun.

The pale beauty of the waterlily has been the subject of many poetic lines from both Eastern and Western traditions. Bhasa, a Sanskrit playwright of the fourth century, wrote in "The Dream Queen" (translated by A. G. Shirreff and P. Lall): "Look, look, my lady, at that line of cranes/In steady flight. 'Tis white and lovely like/A wreath of water lilies." In *The Lady of the Lake* (1810) Sir Walter Scott described the waterlilies among the features of Loch Katrine: "The water-lily to the light/Her chalice reared of silver white." In "Dream-Land," Edgar Allan Poe envisioned

> Lakes that endlessly outspread
> Their lone waters—lone and dead,—
> Their still waters—still and chilly
> With the snows of the lolling lily.

L. E. L. wrote of "The water lilies, that glide so pale,/As if with constant care/Of the treasures which they bear." Oliver Wendell Holmes

wrote of a love affair between a lily and a star in "The Star and the
Water-lily." Having turned down the rose's offer of a summer fling,
the waterlily considers the one who truly captures her heart, the star.
Still wary, she says of the star that there was not a leaf

> That he has not cheered with his fickle smile
> And warmed with his faithless beam—
> And will he be true to a pallid flower,
> That floats on the quiet stream?

Writers have also noted the long-lasting flower of the waterlily.
American gardener Celia Thaxter wrote of a golden day in "August"
—while swallows chattered in their flights overhead,

> Buttercup nodded and said good-bye,
> Clover and daisy went off together,
> But the fragrant water lilies lie
> Yet moored in the golden August weather.

And John Clare wrote in one of several poems titled "Autumn" (1835)
of a scene with waterlilies and of a dragonfly feeding in the sun:

> And meadow pools, torn wide by lawless floods,
> Where water-lilies spread their oily leaves,
> On which, as wont, the fly
> Oft battens in the sun.

In the language of flowers, the Victorians ascribed purity of heart
to the waterlily, in keeping with the ancient symbolism of the lotus.
In art, the waterlilies (*Nymphaea*) of Monet are among the most recognized paintings in the twentieth century.

WEEDS

Whose virtues have not yet been discovered

In 1984 the Weed Science Society of America published a list of 1934 species of plants from fifty-seven botanical families that were deemed to be weeds. The society defines weeds as plants that are in some way pestiferous or bothersome, or that interfere with human activities, particularly the growing of food and agricultural crops; the production of timber, fibers, and dyes; and the yielding of chemical and medicinal products.

The grass family, Poaceae, and the aster family, Compositae (Asteraceae), together contribute a little more than one-third of the most noxious weeds. The sedge family, Cyperaceae, accounts for about 6 percent of the most pesky and most scorned plants. These three families provide nearly 50 percent of the world's worst weeds, despite the fact that they also contribute some of the most important, highly valued plants. Members of the Poaceae, for example, include wheat, rice, corn, oats, and sugarcane.

The term weed is from the Old English *weod* and Anglo-Saxon *wiod*, a general all-purpose word for herbs, grass, or weeds. Comparable terms include bane, meaning a plant capable of destroying life, such as dogbane and wolfbane, and wort, an Old English name for a plant used as food or medicine. The term wort later became a general word for any plant, root, or herb, such as milkwort and lousewort.

"Weed" is not a botanical designation. Which plants are weeds depends on one's viewpoint. Ralph Waldo Emerson delivered a speech, "The Fortune of the Republic," in 1878 at Old South Church in Boston; he said the following about weeds:

> Our modern wealth stands on a few staples, and the interest

nations took in our war was exasperated by the importance of cotton trade. And what is cotton? One plant out of some two hundred thousand known to the botanist, vastly the larger part of which are reckoned weeds. What is a weed? A plant whose virtues have not yet been discovered, —every one of the two hundred thousand probably yet to be of utility in the arts. As Bacchus of the vine, Ceres of the wheat, as Arkwright and Whitney were the demi-gods of cotton, so prolific Time will yet bring an inventor to every plant. There is not a property in nature but a mind is born to seek and find it.

In general, a weed is a plant growing where it is not desired, a plant that is unwanted and generally objectionable. The native American pokeweed, *Phytolacca americana*, graceful and erect in a border of tall perennials, may be desirable and much admired; but the same plant adrift in a rock-strewn bed of *Delosperma* is a weed, pure and simple.

Emerson's fellow transcendentalist, Henry David Thoreau compared weeds with wasted time in "I Am a Parcel of Vain Strivings Tied" (sometimes titled "Sic Vita"):

> A nosegay which Time clutched from out
> Those fair Elysian fields,
> With weeds and broken stems, in haste,
> Doth make the rabble rout
> That waste
> The day he yields.

Weeds have moved authors to include them among their botanical conceits. For example, Shakespeare made several references to weeds. In *Love's Labour's Lost* (1598), he noted, "He weeds the corn and still lets grow the weeding." And later, "To weed this wormwood from your fruitful brain," refers to the still common act of removing weeds from the land allotted for desired plants. Shakespeare even included a bit of gardening advice in *Henry VI, Part II:* "Now 'tis the spring, and weeds are shallow-rooted:/Suffer them now, and they'll o'ergrow the garden/And choke the herbs for want of husbandry."

John Clare allowed even a weed its springtime due in "April," written between 1842 and 1864 at Northampton Asylum:

> 'Tis spring the April of the year,
> The holiday of birds and flowers.
> Some build ere yet the leaves appear,
> While others wait for safer hours:
> Hid in green leaves they shun the shower.
> They're safe and happy all along;
> The meanest weed now finds a flower;
> The simplest bird will learn a song.

Weeds, both generic and specific, have inspired poets to give them positive qualities. American poet Gertrude Hall in "To A Weed" chose even to identify with them:

> You bold thing! thrusting, 'neath the very nose
> Of her fastidious majesty, the rose,
> Even in the best ordainëd garden-bed,
> Unauthorized, your smiling little head!

> The gardener,—mind,—will come in his big boots,
> And drag you by your rebellious roots,
> And cast you forth to shrivel in the sun,
> Your daring quelled, your little weed's life done.

> And when the noon cools, and the sun drops low,
> He'll come again with his big wheelbarrow,
> And trundle you—I don't know clearly where,—
> But *off*, outside the dew, the light, the air.

> Meantime—ah, yes! the air is very blue,
> And gold the light, and diamond the dew,—

You laugh and curtsey in your worthless way,
And you are gay,—so exceeding gay!

You argue in your manner of a weed,
You did not make yourself grow from a seed,
You fancy you've a claim to standing-room,
You dream yourself a right to breath and bloom.

The sun loves you, you think, just as the rose,
He never scorned you for a weed,—*he knows!*
The green-gold flies rest on you and are glad,
It's only cross old gardeners find you bad.

You know, you weed, *I quite agree with you,*
I am a weed myself, and I laugh too,—
Both, just as long as we can shun his eye,
Let's sniff at the old gardener trudging by!

Florence Earle Coates wrote "Jewel-Weed," of the jewelweed or touch-me-not. The weed's botanical name is *Impatiens*, which is the Latin word for impatient, an allusion to the somewhat explosive burst that disperses seed from the plant's ripe seed pods when touched. Coates wrote of a quiet road adorned with impatiens:

Thou lonely, dew-wet mountain road,
Traversed by toiling feet each day,
What rare enchantment maketh thee
Appear so gay?

The sentinels, on either hand
Rise tamarack, birch, and balsam-fir,
O'er the familiar shrubs that greet
The wayfarer;

But here's a magic cometh new—
A joy to gladden thee, indeed:

This passionate out-flowering of
The jewel-weed,

That now, when days are growing drear,
As Summer dreams that she is old,
Hangs out a myriad of pleasure-bells
Of mottled gold!

Thine only, these, thou lonely road!
Though hands that take, and naught restore,
Rob thee of other treasured things,
Thine these are, for

A fairy, cradled in each bloom,
To all who pass the charmèd spot
Whispers in warning: "Friend, admire,—
But touch me not!

"Leave me to blossom where I sprung,
A joy untarnished shall I seem;
Pluck me, and you dispel the charm
And blur the dream!"

It is difficult to complain about grasses and other weeds strangling a garden, even one that has been untended for only a few days, after one reads "The Prairies" by American poet William Cullen Bryant. He called these once vast lands "the gardens of the desert":

As over the verdant waste I guide my steed.
Among the high rank grass that sweeps his sides
The hollow beating of his footsteps seems
A sacrilegious sound.

WISTERIA

When will be mine this lovely flower

Wisteria is a member of the Papilionoideae within the Leguminosae, the pea family. Its six species of woody, deciduous climbing vines are native to Japan, China, and the eastern United States. The white, pink, or blue flowers sit on pendulous racemes that are stunningly showy. The Japanese wisteria is *W. floribunda*, meaning "profusely flowering," and the Chinese wisteria is *W. sinensis*, meaning of China. A garden-originated hybrid of these two species is *W. ×formosa*. An American wisteria *W. frutescens*, meaning shrubby or bushy, was collected by Mark Catesby when he traveled in the American South in the early 1700s. When the wisteria arrived in England, it was called the Carolina kidney bean; it and the Asian wisterias were initially cataloged as soybeans (*Glycine*).

Wisteria is named after Dr. Caspar Wistar, a botanist and professor of medicine at the University of Pennsylvania, who succeeded Thomas Jefferson as the chair of the American Philosophical Society. The plant was named by Thomas Nuttall, a Yorkshireman who moved to Philadelphia and later became a curator of the herbarium at Harvard College. Nuttall had intended to write "Wistaria" in his description of the plant, but a slip of the pen indelibly imprinted the name "Wisteria," which has been in use since 1818. Later, in 1825, the Chinese species classified as *Glycine* were combined under Nuttall's nomenclature. In literature both spellings are used.

In the Japanese language of flowers, the wisteria symbolized youth and poetry, and though it was a flower of April, it became the emblem of summer. In the Victorian language of flowers, the wisteria spoke both "I cling to thee" and "Welcome, fair stranger." Japanese folklore claimed that wisteria roots are nourished if sake is poured at the base of the plant; therefore, visitors to Japanese gardens where wisteria grew were once urged to empty their cups to produce finer blossoms.

In Japanese literature, Princess Wistaria appears in *Genji Monogatari* (eleventh century CE) by Murasaki Shikibu, the pseudonym of an otherwise anonymous female author. The work is regarded as the greatest piece of Japanese fiction. It is the story of court life and the life and loves of Prince Genji, a talented and dashing nobleman. As many flower names are often applied to females, one of the heroine's in *Genji Monogatari* is Princess Wistaria. Prince Genji sings a short ballad comparing the princess and the flower:

> When will be mine this lovely flower
> Of tender grace and purple hue?
> Like the wistaria of the bower,
> Its charms are lovely to my view.

The *Japanese Floral Calendar* (1911) by Ernest W. Clement presents snatches of poems by Japanese writers praising the wisteria. One anonymous poet found the plant visually irresistible:

> O lovely wistaria, now in bloom—
> Twine thy twigs, even though broken,
> To those people who pass by thee,
> Without stopping to admire thy beauty.
>
> Men dare not pass away without looking
> At the wistaria, in a wave of beauty,

Though my small garden be humble,
With nothing attractive for the eye.

And another anonymous poet painted the wisteria into a broader view:

In blossom the wistaria-trees to-day
Break forth that sweep the wavelets of my lake:
When will the mountain cuckoo come and make
The garden vocal with his first sweet lay?

Another poem from the *Japanese Floral Calendar* portrays wisteria as a gift: "What, though I be outside the ring-fence and cannot sit beneath thy shade, thou sendest gentle Wistaria, thy fragrance across it to me, treating me like a friend."

Surrey gardener Mrs. C. W. Earle maintained a diary in which she recorded extensive thoughts and observations of her gardening and cooking. They were published in 1898 as *Pot-Pourri from a Surrey Garden.* Her "notes," as she called them, include a lovely paean to the wisteria:

You have no idea of the joy that is to be got out of a Wistaria, with its beautiful lilac blooms hanging from the bare and twisted branches above your head and the blue sky behind them. The whole effect is indeed different and very superior to that of seeing the blooms hanging straight and flat from branches nailed close to the wall.

YARROW

Sweet yarrow, the first that I have found

Yarrow or milfoil is a member of the Compositae (Asteraceae) that includes about eighty-five species in the genus *Achillea*. The plants are somewhat aromatic, and many species have finely dissected, fern-like leaves. They grow naturally in northern temperate regions. Probably the most widely known and garden-grown species is *A. millefolium*, which has numerous cultivars including 'Lilac Beauty', 'Paprika', and 'Moonshine'. Other garden species include *A. filipendulina* and *A. ptarmica*, or sneeze-wort, which was formerly used as a snuff.

Linnaeus picked the Latin name of the plant because it honors Achilles, legendary warrior and hero of the Trojan War. When Achilles was an infant, his mother, Thetis, dipped him in the River Styx to render him invulnerable. During his immersion, he was entirely covered except for one heel, by which Thetis held him. The heel later proved vulnerable to Paris's arrow, which was un-leashed by Apollo as Achilles fought at the Scaean gate of Troy. Upon Achilles's death there was great wailing by Thetis and her accompanying nereids, who took his ashes to the underworld. In some versions of the myth, Odysseus saw him there strolling among fields of asphodel and lamenting his fate. Because of the plant's association with Achilles and the Trojan War, in the language of flowers the achillea represented war.

While still a youth, Achilles was entrusted by his mother to the teacher Chiron the Centaur, half-horse and half-man. Chiron placed the boy on a diet of bear entrails and honey and taught him music, riding, hunting, the arts, and the healing benefit of plants. Achilles used a salve of yarrow to soothe the wounds of soldiers during the Trojan War. Hence, some old medieval names for the plant included soldier's wound wort, knighten-milfoil, and staunchgrass. Likewise, an old name used by the French was *l'herbe aux charpentiers,* or carpenters' herb, because it was a handy first aid to administer for cuts and nicks incurred in a craftsman's daily work. Natives of western North America were introduced to *Achillea millefolium* by European colonists—they quickly adopted it for soothing bruises and burns.

The origin of the common name yarrow is uncertain; it appears to come from *gearwe* or *gierwan* of Anglo-Saxon origin, which loosely mean "to be ready" or "to prepare," suggesting the yarrow's use against wounds, lovesickness, and various other ills. An old name for yarrow was nose bleed, because the feathery-leaved plant was used either to staunch a nose bleed or to induce one to relieve a migraine. An old superstition attributed to the Suffolk area of England says,

> Green yarrow, green yarrow, wear a white blow,
> If my love loves me, my nose will bleed now!
> If my love don't love me, it 'on't bleed a drop;
> If my love do love me, 'twill bleed every drop.

According to another English superstition, girls could also use the yarrow to assist in divinations of love. A girl was to pluck the leaves from a graveyard while repeating,

> Yarrow, sweet yarrow, the first that I have found,
> In the name of sweet Jesus Christ I pluck thee from the
>   ground;
> As Joseph loved sweet Mary and took her for his dear,
> So in a dream this very night, I hope my love will appear.

After sleeping overnight with the yarrow under her pillow, she was to recite:

> Good morning, good morning, good Yarrow
> And thrice good morning to thee
> Tell me, before this time tomorrow
> Who my true love is to be.

Her dreams the following night would reveal her future husband.

Nicholas Culpeper, author and apothecary, in *The English Physician Enlarged, or the Herbal* (1653) described the virtues of yarrow in the field of animal husbandry:

> Some writers of credit take the pains to inform us what plants cattle will not eat; they judge of this by looking at what are left in the grounds, where they feed; and all such they direct to be rooted up. We have in this an instance, that more care is needful than men commonly take to show what is and what is not valuable. Yarrow is a plant left standing always in fed pasture; for cattle will not eat its dry stalk, nor have the leaves any great virtue after this rises; but yarrow still is useful. It should be sown on barren grass ground, and while the leaves are tender, the cows and horses will eat it heartily. Nothing is more wholesome for them, and it doubles the natural produce. On cutting down the stalks as they rise, it keeps the leaf fresh, and they will eat it as it grows.

On the many virtues of yarrow, the final word appears to be that of John Gerard. He cited the use of a balm of yarrow to cure sexual ailments in young men:

> This hath been prouved by a certain friend of mine, sometimes a Fellow of Kings College in Cambridge, who lightly

brushed the leaves of common yarrow, with hogs-grease, and applied warme into the privie parts, and thereby did divers times helpe himselfe, and others of his fellowes, when he was a student and a single man living in Cambridge.

ZINNIA

Like the visions fresh and bright

The eleven species of *Zinnia* are members of the Compositae (Asteraceae), and the majority have their origin in Mexico. Some species are found in the southwestern United States and Argentina, however. While the cultivated zinnia's wild ancestors are believed to have been a pale lilac color, perhaps *Z. elegans*, many current garden forms are brightly colored hues of red, yellow, orange, and purple.

Zinnias were named by Linnaeus for a German professor of botany, medicine, and physics, Johann Gottfried Zinn, who taught at the University of Göttingen. He was the author of *Hortus Göttingensis*, a flora published around 1750. Zinnias were known as the Brazilian marigold when Linnaeus cataloged them with the mistaken belief that they came from South America.

When the Spanish arrived in Mexico in the 1520s, they found zinnias growing in the gardens of Montezuma, and they called the unattractive wild zinnia *mal de ojos*, meaning "ugly to the eye" or eyesore. The Aztecs grew *Zinnia angustifolia*, meaning narrow-leaved, and *Z. elegans*, elegant, despite what the Spanish thought of its beauty. Seed from zinnias was sent to Spain from Mexico in the late 1700s by Professor Casimir Gomez de Ortego to his Lady Bute, the wife of the British ambassador to Madrid.

From Madrid, they reached Kew through the hands of her father-in-law, John Stuart, Third Earl of Bute, the director at the Royal Botanic Gardens.

Among Native Americans of the southwestern United States, particularly the Navajo and the Apache, the zinnia was known as Ever-Changing Woman, the personification of Mother Nature. She was the goddess who would grow old and then young again; the zinnia shared this trait, with its characteristic aging of the ray flowers and subsequent appearance of the young disk flowers. For this habit, it is also sometimes called youth-and-old-age plant.

A Navajo legend describes a tribe whose crops were being destroyed by drought and insects. The tribe sought advice from Ever-Changing Woman, who advised them to seek the counsel of the Spider Woman. The tribe sent Straight Arrow, the twelve-year-old son of the chief, to find the solution to its problems. As he sought the Spider Woman, sulfur-colored zinnia flowers appeared along his trail. When he found her, she said, "Your tribe's problem is the work of the Chindi [the devil] who must be exorcised from the fields." She told them to plant at night the sulfur-colored flowers that Ever-Changing Woman had caused to sprout along Straight Arrow's trail. Ever since, the Navajo have planted zinnias in their cornfields and have never harmed a spider or its web on a stand of zinnias or in a cornfield.

The following quatrain by an anonymous author, while not entirely horticulturally correct as to the plant's growing conditions, suggests the origin of the zinnia's symbolism in the Victorian language of flowers, which referred to fond thoughts of absent friends:

> The zinnia's solitary flower
> Which blooms in forests lone and deep
> Are like the visions fresh and bright
> That faithful, absent hearts will keep.

Walter Pater, English essayist and critic of the nineteenth century, drew upon the connection between zinnias and friendship when he wrote his masterwork *Marius the Epicurean* (1885), a philosophical romance that describes the young manhood of the Roman Marius. With his friend Favian, "They visited the flower-market, lingering where the *coronarii* [the vendors and makers of crowns and garlands] pressed on them the newest species, and purchased zinnias, now in blossom (like painted flowers, thought Marius), to decorate the folds of their togas." Although Pater's work is generally praised, his horticulture was lacking: zinnias were not known to ancient Romans.

In the southern United States of the mid-1900s, zinnias were the definitive summer annual, grown at a cost of a few pennies. The small packets were usually obtained by mail order and the seed loosely scattered on sand and clay, where they provided radiant hues in the summer sun until the frost cut their bloom.

CALENDAR OF FLOWERS

OTH GARDENS and books must have boundaries, places where they begin and end. Plants offer such an endless array of possible subjects for a garden that every gardener must face constraints to limit the collection. Likewise, this collection must be limited, leaving many plants and literary sources untouched.

The works cited in this collection of garden plants in literature have come from both learned and ordinary people, many of whom were the source of legends and poems that were later recorded by others. Their voices have helped to characterize the garden plants in this collection. Thus, it is fitting to close with "An Early Church Calendar of English Flowers," a poetic chronology of the bloom times for garden flowers, which are familiar to so many gardeners. It relates the identification of flowers to church festivals and holy personages. Unfortunately the piece has no identifiable author or clear date of composition.

> The Snowdrop, in purest white arraie,
> First rears her hedde on Candlemas daie;
> While the Crocus hastens to the shrine
> Of Primrose lone on Saint Valentine.
> Then comes the Daffodil, beside
> Our Ladye's Smock at our Ladye-tide.
> Aboute Saint George, when blue is worn,
> The blue Harebells the fields adorn;
> Against the day of Holie Cross,
> The Crowfoot gilds the flowerie grasse.

When Saint Barnabie bright smiles night and daie,
Poor Ragged Robin blossoms in the haie.
The Scarlet Lychnis, the garden's pride,
Flames at Saint John the Baptist's tide.
From Visitation to Saint Swithin's showers,
The Lillie White reigns Queen of the floures;
And Poppies, a sanguine mantle spred
For the blood of the Dragon Saint Margaret shed.
Then under the wanton Rose, agen,
That blushes for Penitent Magdalen,
Till Lammas daie, called August's Wheel,
When the long Corn stinks of Cammomile.
When Mary left us here belowe,
The Virgin's Bower is full in blowe;
And yet anon, the full Sunflowre blew,
And became a starre for Bartholomew.
The Passion-floure long has blowed,
To betoken us signs of the Holy Roode.
The Michaelmas Daisy, among dede weeds,
Blooms for Saint Michael's valourous deeds;
And seems the last of floures that strode,
Till the feste of Saint Simon and Saint Jude—
Save Mushrooms, and the Fungus race,
That grow till All-Hallow-tide takes place.
Soon the evergreen Laurel alone is greene,
When Catherine crownes all learned menne,
The Ivie and Holly Berries are seen,
And Yule Log and Wassaile come round agen.

SAINTS OF THE SPADE

LTHOUGH heroes and patron saints are a part of the heritage and history of most western Christian nations, it is more often military leaders whose legends, stories, and myths grow larger than life and become the subjects of great admiration. Veneration for the saints, on the other hand, often fades as their stories become more and more otherworldly. Their decrescendo into the supernatural usually begins by their having died in a horrible but beatific manner. But saints were flesh-and-blood humans, not angels. They were venerated because they perhaps eccentrically took the Bible and its gospels literally in their lives. They could be virtuous and pious and above all obsessed and driven. They aspired to a lofty, noble plane, and upon death, it is believed, they were given the ability to intercede on the behalf of those on earth.

A few saints are associated with gardens and gardening. They have been depicted in tapestries, mosaics, miniatures, paintings, and various writings. Factual details regarding the histories of these saints vary from writer to writer within hagiographic circles, those studying the lives of saints. Doubtless their glories have been embellished and altered through the crossing of geographic, linguistic, and religious boundaries. At times, the legends associated with their lives stretch credulity.

St. Phocas of Sinope

The earliest patron saint of gardening appears to be St. Phocas, who lived in the third century outside the city of Sinope near the Black Sea in Pontus, which is now Turkey. Phocas dedicated his life to growing vegetables for the poor of his district and flowers for himself on the little plot of land available to him. When he was not growing plants, which he was skilled at, he spent his time in prayer. He often provided hospice for travelers on the Black Sea.

One day, during an era of religious persecution of Christians, two soldiers came to St. Phocas and he welcomed them with his famous hospitality, offering them food and lodging for the night. After dinner they told him they were searching for the Christian named Phocas and that they had orders to kill him. Phocas did not reveal his identity but said he knew the man and would show him to them the next day. That night after his prayers, Phocas went to his garden and dug a grave. In the morning as the strangers were about to depart he led them to his garden and told them that they should carry out their orders, saying, "I myself am the man." He assured them he regarded martyrdom as the greatest gift they could give him and urged them to proceed with their orders. The soldiers cut off his head and buried him among the flowers and plants he had tended.

Today in Italy, St. Phocas is depicted on the cathedral of Palermo and in the mosaics in St. Mark's Cathedral in Venice. His feast day is celebrated on various dates, most commonly 22 September, and his emblem is a spade or shovel.

St. Fiacre of France

St. Fiacre was an Irish hermit-abbot who lived in the seventh century. He began his career as an ordained priest in Ireland and founded a hermitage in County Kilkenny. In 628, he left Ireland to preach to the Gauls near Meaux in Brittany, France. He was welcomed by St. Faro and lived as an anchorite in the great forests, where he made a garden among the wolves and wild boars, none of which harmed him.

In the legend, the local bishop offered St. Fiacre as much land as he could plow in a day, and he succeeded in clearing several acres for his claim. He built an oratory and later a monastery while living a life of self-denial. At Meaux he attracted many who sought spiritual guidance. He planted an extensive vegetable garden and became a gifted horticulturist. According to legend, his garden was "miraculously enclosed," which a woman told the bishop was the work of the Evil One. The Bishop visited the garden in the middle of the forest and found that the woman's claim was false. After that, no woman would go to St. Fiacre's oratory for fear of being struck with blindness. The story of the forest garden is remarkably similar to the legend of the Christmas rose told from Scandinavian folklore by Swedish author Selma Lagerlöf (see the chapter "Hellebore").

One legend of St. Fiacre says he was so despondent over the dispute with the woman about his garden that he sat down heavily on a stone leaving an imprint of his buttocks. The stone, later moved to St. Fiacre-en-Brie, became a shrine for pilgrims to sit upon to cure piles. Additional legends say a noblewoman who visited St. Fiacre's chapel in the 1500s instantly became insane. Even as late as the 1700s no woman would enter the chapel named for St. Fiacre at the Cathedral of Meaux. Anne of Austria, queen to Louis XIII of France, declined to enter the chapel during a visit in 1648. Nevertheless, she credited St. Fiacre with the safe delivery of her son, Louis XIV, the Sun King.

St. Fiacre is depicted with a spade or shovel in one hand and either flowers or a book in the other. In a French book of hours, a miniature of St. Fiacre shows him holding a spade with the Cathedral of Meaux in the background. Framing the miniature are flowers, leaves, and butterflies. Saint Fiacre died on 12 August 670. His feast day is celebrated by some on 30 August and by others on 1 September. There is record of an old alpine strawberry named after him.

The French word for a taxi or cab as well as for a hackney cab, a small horse-drawn carriage, is *le fiacre*. This name comes from the Hotel St. Fiacre where hackney cabs were first available for hire in about 1690. As a result, St. Fiacre is also the patron saint of French taxi drivers.

Santa Rosa de Santa Maria, The Rose of Lima

Spaniards brought Christianity to most of South America. During the era of the conquistadors, the Spanish established both religious and colonial authority over the indigenous peoples. With that authority came both military heroes and saints from the New World. One of these saints was Santa Rosa, the preeminent saint of Hispanic America.

In the late 1550s, the cathedral at Lima, Peru, was the site of a ceremony honoring the Virgin Mary, during which the archbishop reverently placed a rose at the feet of the statue of the Virgin. This rose was the first grown from seed introduced into Peru in 1552. Some of the seed had been planted on the grounds at the Hospital del Espíritu Santo and some on nearby lands that were tended by the de Flores family. In 1586 a daughter was born to the de Flores family, and she played among the roses in their garden. She was christened Isabel after her grandmother but historically became known as Santa Rosa de Santa Maria and eventually Santa Rosa de Lima.

Although of mixed Hispanic, Incan, and Creole descent, she had golden hair and a face of rare beauty—"*una rosa perfecta*" was uttered by the house maid upon seeing the baby for the first time. As a child she was spontaneously able to read and write and play musical instruments without prior training or lessons. Rosa vowed at age five to be the wife of Christ and began subjecting herself to ferocious and extraordinary torments. She frequently lived on bread and water and occasionally a homemade brew of sheep's gall and bitter herbs. She once fasted for fifty days. Routinely she would sleep only two hours per day and even then only on a bed of broken kitchenware, stones, and wood scraps. While in bed she would sleep sitting up and tie her hair to nails on the wall to prevent herself from oversleeping the allotted two hours. Santa Rosa wore a crown of roses that hid from view thirty-three brow-piercing spikes of metal, one for each year of the life of Christ. To destroy the beauty of her hands, she rubbed lime and red peppers on them, then wore gloves of nettles. When she worked in the family garden, which provided income primarily from growing flowers, she dragged a heavy cross. But Rosa was given a green thumb. She experimented with topiary—she trimmed rosemary bushes to look like Calvary—and she grew sweet basil for the adornment of the chapel of the rosary. In the garden she kept a small stone cell for quiet seclusion for her daily twelve-hour prayer sessions.

She died on 23 August 1617 of fever and pains that led to a stroke and paralysis on the left side of her body. Santa Rosa was not given the benefit of a drink of water during her dying, for the doctors thought water would redouble her fever. She was honored with a public funeral and in 1671 became the first New World native to be canonized. She is buried at St. Dominic's Church in Lima under the altar in the crypt. The Rose of Lima is the patron saint of florists and gardeners. She is typically depicted wearing a crown of roses, and her feast day is 23 August.

St. Swithin

St. Swithin's Day, if thou dost rain,
For forty days it will remain;
St. Swithin's Day, if thou be fair,
For forty days 'twill rain nae mair.

In the Anglican church, 15 July is the festival day for St. Swithin, who in popular lore foretells the weather of the following forty days. Swithin was bishop of Winchester and a counselor to Egbert, King of the West Saxons, who had helped elect him bishop in October of the year 852. Swithin performed numerous miracles, cures, and good deeds as bishop. Among the miracles was the mending of a basket of broken eggs belonging to a poor peasant woman who had been tripped by an impish boy on the cobblestone bridge at Winchester. Swithin died on 2 July 862 and, as he had asked, he was buried not in the chancel, but outside where the "feet of passersby might tread and the sweet rain of heaven fall" on his tomb. He was canonized in 971, and his bones were dug up on 15 July and placed inside the cathedral as traditionally befitted a saint. In protest of the disregard for his burial wishes, St.

Swithin arranged for a forty-day deluge of rain. His body was moved a second time in 1094 to a newer cathedral.

St. Swithin is a patron saint invoked both for and against rain. When it rains on 15 July, people say that St. Swithin is christening the apples. The flower traditionally associated with him is a marigold.

A Garden Variety of Saints

A well-known saint much associated with gardening is St. John the Baptist. Gardeners are familiar with the plant commonly called St. John's wort, *Hypericum*, which is generally blooming on 24 June, the feast day of St. John, which celebrates his nativity. Though astronomically not the summer solstice, 24 June was long ago determined to be midsummer's day. On this day the plant was traditionally placed above icons to ward off evil and bad spirits. In revenge, the devil is said to have pricked holes in *H. perforatum*, which has oil glands that appear as clear spots in its leaves when held up to the light. The flowers of the wort, gathered after sunset on St. John's eve, were widely used during the pagan midsummer festival. The celebrations included bonfires that kept away witches and demons by extending the light of the sun. After St. John was beheaded, 24 June became known as St. John's Day, for Christians considered him the herald of light. The date honored for his beheading is 29 August, on which day the yellow hollyhock decorates churches in his memory.

One early saint could be called a closet gardener. He was St. Maurille of France, called St. Maurilius in Italian, who died ca. 430 CE. Originally from Milan, though he followed St. Martin to Tours and was ordained there, St. Maurille was later made bishop after he successfully prayed for fire to come down from heaven and destroy heathens and their idols. He felt the honor given him was too great, and he sailed to Roman-occupied England to escape his obligations. There he changed his guise to that of a fifth-century gardener. He was recognized, however, and sent back to Gaul to fulfill his duties as bishop. Saint Maurille is depicted on a fifteenth-century tapestry in Angers

Cathedral in Maine-et-Loire, France, working in the garden with a long-handled spade, with fruit trees and gooseberry bushes behind him and flowers before him.

A patron of gardeners is St. Adelard of France, who died in 827. He was a cousin of Charlemagne. Adelard went to a monastery at the age of twenty, where his first duties were weeding and digging in the garden. He became an advisor to the political rulers of his district but was banished to Normandy for the advice he gave them. He remained in exile there, occupying himself as a gardener. His feast day is 2 January.

St. Gertrude of Nivelles was born in 626. Invoking St. Gertrude might be of use today in repelling moles and voles from the garden, because in life she was able to vanquish plagues of field mice. She is often pictured with mice at her feet. By tradition, fine weather on her feast day of 17 March, which she shares with St. Patrick, is a good omen and marks the beginning of the outdoor gardening season.

The sixth day of February is the feast of St. Dorothy of Italy, who survived being thrown into boiling animal fat, but who was beheaded ca. 313 CE. She returned to earth as an angel child, bringing three apples and three roses from the garden of paradise to the lawyer and cynic Theophilus, who had mocked her on her way to death. He was astonished at the sight and was instantly converted by the miracle. Theophilus was himself later beheaded, cut up into little pieces, and fed to the birds. Saint Dorothy is depicted with apples, either in her hand or in a basket, and with roses.

Another saint of the spade was Teresa of Lisieux, who was a Carmelite sister born in 1873. She became known as the Little Flower. Teresa endured tedious miseries, such as not scratching an itch and bearing with benign dignity any insect crawling on her skin. Reminiscent of the bizarre behavior of Rose of Lima, she slept under a thickness of blankets in the hot French summers and with no cover in the cold winters. As a child she was full of fun, girlish and a practical joker; she was once scolded for pulling up flowers instead of weeds from the garden. Saint Teresa wrote a popular autobiography, *L'histoire d'une âme* (The Story of a Soul), in which she said, "After my death, I will let fall showers of roses." She died of tuberculosis in 1897. Her feast day is 1 October, and she is typically shown in art wearing a Carmelite habit and holding a bunch of roses.

The patron saint of florists is Rose of Viterbo. She lived in the mid-thirteenth century and was a child saint, living to be only seventeen. Despite living in exile for making anti-crown predictions, at the age of ten she played with a friend of the emperor. Once, she and the boy got into an argument that led to a challenge in which Rose dared her playmate to jump into a bonfire. The two of them did, but only Rose emerged unharmed, winning the argument. Saint Rose of Viterbo's feast day is 4 September.

Finally, a potpourri of saints cannot overlook the original patrons of gardeners, Adam and Eve, who unwittingly found themselves tending the Garden of Eden. They are the first recorded pomoculturalists (fruit growers). Adam and Eve are generally considered saints of the Old Testament, because it is said they were raised from hell sometime between Good Friday and the Resurrection. Old Adam's feast day is 24 December.

Much gardening help could be invoked from the saints: St. Urban of Langres will guard against blight and frost, St. Magnus of Fussen will ward off caterpillars, St. Genevieve will prevent drought, St. Swithin will bring or not bring rain, St. Dominic of Silos will overcome plagues of insects, and St. Leufredus will kill white flies. If you want to till or plow a straight row in your garden, then St. Isidore of Madrid might help. Legend has it that when he plowed his fields with a team of oxen, a second team of white oxen led by angels plowed along beside him. Saint Valentine, whose signatory flower is the crocus, should be invoked when planting bulbs in the fall. If your green thumb has turned brown and your gardening skills are in need of more help than the above saints can offer, you may want to invoke thoughts of St. Jude, who intercedes in hopeless cases. And on All Saints Day, 1 November, the day after Halloween, you would be wise to put all your garden supplications in one basket.

BIBLIOGRAPHY

Abrams, M. H., ed. 1974. *Norton Anthology of English Literature.* New York: W. W. Norton.

Allison, Alexander W., et al. 1975. *Norton Anthology of Poetry.* rev. ed. New York: W. W. Norton.

Andersen, Hans C. 1988. *Snowdrop.* Trans. Marlee Alex. Waco, Texas: Word.

Angel, Marie. 1985. *A Floral Calendar and Other Flower Lore.* London: Pelham Books.

Arnim-Schlagenthin, Mary Annette. 1900. *Elizabeth and Her German Garden.* London: Macmillan.

Arnold, Matthew. 1897. *Poetical Works of Matthew Arnold.* New York: Thomas Y. Crowell.

Ault, Norman, ed. 1925. *Elizabethan Lyrics from the Original Texts.* New York: Longmans, Green.

Austin, Alfred. 1891. *Lyrical Poems.* London: Macmillan.

Avery, Catherine B. 1962. *New Century Classical Handbook.* New York: Applton-Century-Crofts.

Bacon, Francis. 1936. *Essays of Francis Bacon.* Ed. Clark Sutherland Northrup. Boston: Houghton Mifflin.

Baker, Margaret. 1996. *Discovering the Folklore of Plants.* 3rd ed. Princes Risborough, Buckinghamshire: Shire Publications.

Barnette, Martha. 1992. *A Garden of Words.* New York: Random House.

Bartlett, John. 1980. *Familiar Quotations.* Ed. Emily Morison Beck. Boston: Little, Brown.

Bartram, William. 1980 [1792]. *Travels Through North and South Carolina, Georgia, East and West Florida.* Reprint. Charlottesville, Virginia: University Press of Virginia.

Beals, Katharine M. 1917. *Flower Lore and Legend.* New York: Henry Holt.

Beaumont, Francis, and John Fletcher. 1969. *Works of Francis Beaumont and John Fletcher*. Eds. Arnold Glover and A. R. Waller. New York: Octagon Books.

Behn, Aphra. 1992. *Works of Aphra Behn*, vol. 1. Columbus: Ohio State University Press.

Benét, William Rose. 1987. *Benét's Reader's Encyclopedia*. 3rd ed. New York: Harper and Row.

Blake, William. 1941. *Complete Poetry and Selected Prose of John Donne and the Complete Poetry of William Blake*. New York: Random House.

Bloom, J. Harvey. 1903. *Shakespeare's Garden*. London: Methuen.

Book of 1,000 Poems. 1993. Avenel, New Jersey: Wings Books.

Bowers, Fredson, ed. 1966. *Dramatic Works in the Beaumont and Fletcher Canon*, vol. 3. Cambridge: Cambridge University Press.

Brewer, Cobham. 1993 [1894]. *Dictionary of Phrase and Fable*. Reprint. Ware, Hertfordshire: Wordsworth Editions

Brontë, Anne. 1985 [1848]. *Tenant of Wildfell Hall*. Reprint. London: Penguin Paperbacks.

Brontë, Emily. 1976. *Wuthering Heights*. Oxford: Clarendon Press.

Bronzert, Kathleen, and Bruce Sherwin, eds. 1993. *Glory of the Garden*. New York: Avon Books.

Brooke, Rupert. 1970. *Poetical Works*. Ed. Geoffrey Keynes. London: Faber and Faber.

Brooke, Rupert. 1987. *Rupert Brooke: The Collected Poems*. London: Sidgwick and Jackson.

Brown, Thomas Edward. 1900. *Collected Poems of T. E. Brown*. London: Macmillan.

Browne, Thomas. 1968. *Prose of Sir Thomas Browne*. New York: New York University Press.

Browne, William. 1894. *Poems of William Browne of Tavistock*, vol. 1. London: Lawrence and Bullen.

Browning, Elizabeth Barrett. 1890. *Poetical Works of Elizabeth Barrett Browning*, vol. 1. New York: Worthington.

Browning, Elizabeth Barrett. 1978. *Aurora Leigh and Other Poems*. London: The Women's Press.

Browning, Gareth H. 1952. *Naming of Wild-Flowers*. London: Williams and Northgate.

Browning, Robert. 1895. *Complete Poetic and Dramatic Works of Robert Browning*. Boston: Houghton Mifflin.

Bryant, William Cullen, ed. 1970 [1870]. *Library of World Poetry*. Reprint. New York: Avenel Books.

Bryant, William Cullen. 1901. *Poetical Works of William Cullen Bryant*. New York: D. Appleton.

Buchan, Ursula, and Nigel Colborn. 1987. *Classic Horticulturist*. London: Cassell.

Burns, Robert. 1897. *Complete Poetical Works of Robert Burns*. Eds. W. E. Henley and T. F. Henderson. Cambridge, Massachusetts: Houghton Mifflin.

Burton, Robert. 1903. *Anatomy of Melancholy*, vol. 2. Ed. A. R. Shilleto. London: George Beiland and Sons.

Byron, Lord. 1850. *Poetical Works*. Boston: Phillips, Sampson.

Byron, Lord. 1905. *Complete Poetical Works of Byron*. Cambridge, Massachusetts: Houghton Mifflin.

Byron, Lord. 1936. *Childe Harold's Pilgrimage*. Garden City, New York: Doubleday, Doran.

Camoëns, Luis de. 1940 [1655]. *The Lusiad*. Reprint. Trans. Richard Fanshawe. Ed. Jeremiah D. M. Ford. Cambridge, Massachusetts: Harvard University Press.

Carman, Bliss, ed. 1976. *Oxford Book of American Verse*. Miami: Grainger Books.

Chaucer, Geoffrey. nd. *Student's Chaucer*. Ed. Walter W. Skeat. Oxford: Clarendon Press.

Clare, John. 1935. *Poems of John Clare*. 2 vols. Ed. J. W. Tibble. London: J. M. Dent and Sons.

Clare, John. 1973. *Selected Poems*. Ed. J. W. Tibble. New York: Dutton.

Clement, Ernest W. 1911. *Japanese Floral Calendar*. Chicago: Open Court Publishing.

Clissold, Stephen. 1972. *Saints of South America*. London: Charles Knight.

Coats, Alice M. 1965. *Garden Shrubs and Their Histories*. New York: E. P. Dutton.

Coats, Alice M. 1968. *Flowers and Their Histories.* London: Adam and Charles Black.

Coats, Alice M. 1975. *Treasury of Flowers.* New York: McGraw Hill.

Coats, Peter. 1970. *Flowers in History.* New York: Viking Press.

Coffey, Timothy. 1993. *History and Folklore of North American Wildflowers.* Boston: Houghton Mifflin.

Coleridge, Samuel Taylor. 1912. *Complete Poetical Works of Samuel Taylor Coleridge.* Oxford: Clarendon Press.

Compact Edition of the Oxford English Dictionary. 3 vols. 1971. London: Oxford University Press.

Cook, Eliza. 1856. *Poetical Works of Eliza Cook.* Philadelphia: Willis P. Hazard.

Coombes, Allen J. 1994. *Dictionary of Plant Names.* Portland, Oregon: Timber Press.

Cowper, William. 1905. *Poems of William Cowper.* London: Methuen and Company.

Cox, E. H. M. 1931. *Gardener's Chapbook.* London: Chatto and Windus.

Crabbe, George. 1988. *Complete Poetical Works.* 2 vols. Oxford: Clarendon Press.

Crane, Florence Hedleston. 1932. *Flowers and Folk-Lore from Far Korea.* Tokyo: San Seido.

Creekmore, Hubert, ed. 1952. *Little Treasury of World Poetry.* New York: Charles Scribner's Sons.

Croly, George. 1830. *Poetical Works of the Rev. George Croly.* 2 vols. London: Henry Colburn and Richard Bentley.

Crowell, Robert L. 1982. *Lore and Legends of Flowers.* New York: Thomas Y. Crowell.

Culpeper, Nicholas. 1992 [1826]. *Culpeper Complete Herbal.* Reprint. Leicester, England: Magna Books.

Darwin, Erasmus. 1807. *Botanic Garden, a Poem in Two Parts.* New York: T. and J. Swords.

Day, Sarah J. 1909. *Mayflowers to Mistletoe: A Year with the Flower Folk.* New York: G. P. Putnam's Sons.

De Gex, Jenny. 1994. *Shakespeare's Flowers.* London: Pavilion.

De Gex, Jenny. 1996. *Bible Flowers.* New York: Harmony Books.

Delibes, Léo. 1890. *Lakmé*. Libretto by Gondinet and Ph. Gille. Trans. Theodore T. Barker. Boston: Oliver Ditson.

De Vere, Aubrey. 1884. *Search After Proserpine and Other Poems*, vol. 1. London: Kegan Paul, Trench.

De Vere, Aubrey. 1904. *Poems from the Works of Aubrey De Vere*. London: Catholic Truth Society.

Dickens, Charles. 1943. *David Copperfield*. New York: Dodd, Mead.

Dickens, Charles. 1996. *Mystery of Edwin Drood*. Rutland, Vermont: C. E. Tuttle.

Dowden, Anne Ophelia T. 1963. *Look at a Flower*. New York: Thomas Y. Crowell.

Drabble, Margaret, ed. 1985. *Oxford Companion to English Literature*. 5th ed. New York: Oxford University Press.

Drayton, Michael. 1953. *Poems of Michael Drayton*. Ed. John Buxton. London: Routledge and Kegan Paul Ltd.

Dunbar, Lin. 1989. *Ferns of the Coastal Plain: Their Lore, Legends and Uses*. Columbia, South Carolina: University of South Carolina Press.

Dumas, Alexandre. 1893. *Vicomte de Bragellone, or Ten Years Later*. London: J. M. Dent.

Durant, Mary. 1976. *Who Named the Daisy? Who Named the Rose?* New York: Congdon and Weed.

Earle, Mrs. C. W. 1898. *Pot-Pourri from a Surrey Garden*. London: Smith, Elder.

Earle, Mrs. C. W. 1899. *More Pot-Pourri from a Surrey Garden*. London: Smith, Elder.

Eliot, George. 1948. *Silas Marner*. New York: Dodd, Mead.

Elliott, Ebenezer. 1833. *Splendid Village: Corn Law Rhymes*. London: Benjamin Steill.

Elliott, Ebenezer. 1876. *Poetical Works of Ebenezer Elliott*. 2 vols. Ed. Edwin Elliott. London: Henry S. King.

Elphinstone, Margaret, ed. *Treasury of Garden Verse*. Edinburgh: Canongate.

Emerson, Ralph Waldo. 1904. *Complete Poetical Works of Ralph Waldo Emerson*. Ed. Edward W. Emerson. Boston: Houghton Mifflin.

Emerson, Ralph Waldo. 1906. *Works of Ralph Waldo Emerson*, vol. 11. Ed. Edward W. Emerson. Boston: Houghton Mifflin Fireside Edition.

Evans, Bergen. 1968. *Dictionary of Quotations*. New York: Delacourte Press.

Evelyn, John. 1982 [1699]. *Acetaria: A Discourse of Sallets*. Reprint. London: Prospect Books.

Ewen, David. 1971. *New Encyclopedia of the Opera*. New York: Hill and Wang.

Fanshawe, Richard. 1997. *Poems and Translations of Sir Richard Fanshawe*, vol. 1. Oxford: Clarendon Press.

Farmer, David Hugh. 1992. *Oxford Dictionary of Saints*. Oxford: Oxford University Press.

Felleman, Hazel. 1936. *Best Loved Poems of the American People*. New York: Doubleday.

Ferris, Helen. 1957. *Favorite Poems Old and New*. New York: Doubleday.

Flower Lore. nd. Belfast: McCaw, Stevenson and Orr.

Fogg, H. G. Witham. 1977. *History of Popular Garden Plants from A to Z*. New York: A. S. Barnes.

Folkard, Richard, Jr. 1884. *Plant Lore, Legends and Lyrics*. London: Sampson Low, Marston, Searle and Rivington.

Forster, E. M. 1971. *Room with a View, Howard's End, Maurice*. New York: Quality Paper Back.

Foster, Thomas E., and Elizabeth C. Guthrie, eds. 1995. *Year in Poetry*. New York: Crown.

Fowler, Alastair, ed. 1992. *Seventeenth Century Verse*. Oxford: Oxford University Press.

Friend, Hilderic. 1884. *Flowers and Flower Lore*. 2 vols. London: Swan Sonnenschein.

Fuller, Thomas. 1952. *The Worthies of England*. Ed. John Freeman. London: George Allen and Unwin.

Furness, Mrs. Horace Howard. 1874. *Concordance to Shakespeare's Poems*. Philadelphia: J. B. Lippincott.

Furse, Margaret Celia. 1919. *Gift*. London: Constable.

Gay, John. 1926. *Poetical Works of John Gay*. London: Oxford University Press.

Gerard, John. 1975 [1633]. *Herbal or General History of Plants*. rev. ed. by Thomas Johnson. New York: Dover.

Gilbert, W. S. 1958. *First Night Gilbert and Sullivan*. New York: Heritage Press.

Gordon, Lesley. 1977. *Green Magic: Flowers, Plants and Herbs in Lore and Legend.* New York: Viking.

Gordon, Lesley. 1985. *Mystery and Magic of Trees and Flowers.* London: Grange.

Gounod, Charles. 1905. *Faust: A Lyric Drama in Five Acts.* Libretto by J. Barbier and M. Carré. Philadelphia: Oliver Ditson.

Great Literature (CD-ROM). 1992. Parsippany, New Jersey: Bureau Development.

Greenaway, Kate. 1992 [1884]. *Language of Flowers.* Reprint. New York: Dover.

Greene, David H., ed. 1954. *Anthology of Irish Literature.* New York: Modern Library.

Greene, Ellin. 1990. *Legend of the Christmas Rose: A Retelling of the Original Story by Selma Lagerlöf.* New York: Holiday House.

Grieve, Mrs. M. 1992 [1931]. *Modern Herbal.* Reprint. London: Dorset Press.

Griffiths, Mark. 1994. *Index of Garden Plants.* Portland, Oregon: Timber Press.

Grigson, Geoffrey. 1974. *Dictionary of English Plant Names.* London: Allen Lane.

Habington, William. 1948. *Poems of William Habington.* London: University Press of Liverpool.

Hadfield, Miles, and John Hadfield. 1964. *Gardens of Delight.* Boston: Little, Brown.

Harvey, Gail. 1991. *Poems of Flowers.* New York: Avenel Books.

Hawker, Robert Stephen. 1899. *Poetical Works of Robert Stephen Hawker: Sometime Vicar of Morwenstow, Cornwall.* Ed. Alfred Wallis. London: John Lane.

Healey, B. J. 1972. *Gardener's Guide to Plant Names.* New York: Charles Scribner's Sons.

Herrick, Robert. 1963. *Complete Poetry of Robert Herrick.* New York: New York University Press.

Hibberd, Shirley. 1986. *Amateur's Flower Garden.* Portland, Oregon: Timber Press.

Higham, T. F., and C. M. Bowra. 1938. *Oxford Book of Greek Verse in Translation.* Oxford: Oxford University Press.

Hillier Nursery. 1991. *Hillier Manual of Trees and Shrubs*. Newton Abbot, U.K.: David and Charles.

Hogg, James. 1970. *Selected Poems*. Oxford: Clarendon Press.

Hollander, John, ed. 1996. *Garden Poems*. New York: Alfred A. Knopf.

Hollingsworth, Buckner. 1958. *Flower Chronicles*. New Brunswick, New Jersey: Rutgers University Press.

Holmes, Oliver Wendell. 1895. *Complete Poetical Works of Oliver Wendell Holmes*. Ed. Horace E. Scudder. Boston: Houghton Mifflin.

Holy Bible (KJV). nd. Cleveland: World Publishing.

Hood, Thomas. 1866. *Poetical Works of Thomas Hood*. New York: G. P. Putnam, Hurd and Houghton.

Hottes, Alfred Carl. 1949. *Garden Facts and Fancies*. New York: Dodd, Mead.

Howitt, Mary Botham. 1847. *Poetical Works of Howitt, Milman, and Keats*. Philadelphia: Crissy and Markley.

Howitt, Mary Botham, ed. 1854. *Pictorial Calendar of the Seasons*. London: Henry G. Bohn.

Howitt, William. 1827. *Desolation of Eyam, the Emigrant and Other Poems*. London: Wightman and Cramp.

Hunt, John Dixon, ed. 1993. *Oxford Book of Garden Verse*. Oxford: Oxford University Press.

Hunt, Leigh. 1849. *Poetical Works of Leigh Hunt*. London: Edward Moxon.

Hunt, Violet. 1897. *Unkist, Unkind!* London: Chapman and Hall.

Hyam, Roger, and Richard Pankhurst. 1995. *Plants and Their Names: A Concise Dictionary*. New York: Oxford University Press.

Ingelow, Jean. 1863. *Poems*. London: Longman, Green, Longman, Roberts and Green.

Ingram, J. H. 1869. *Flora Symbolica or the Language and Sentiment of Flowers*. London: Frederick Warne.

Jefferies, Richard. 1879. *Wild Life in a Southern County*. London: Smith, Elder.

Jefferies, Richard. 1885. *Open Air*. London: Chatto and Windus.

Johnson, Samuel, ed. 1810. *Works of the English Poets from Chaucer to Cowper*, vols. 11 and 16. London: C. Whittingham.

Jonson, Ben. 1854. *Works of Ben Jonson*. Boston: Phillips, Sampson.

Jonson, Ben. 1938. *Selected Works*. New York: Random House.

Jonson, Ben. 1947. *Poems [and] The Prose Works*. Eds. C. H. Herford, Percy and Evelyn Simpson. Oxford: Clarendon Press.

Joy, Margaret. 1984. *Days, Weeks, and Months*. London: Faber and Faber.

Keats, John, and Percy Bysshe Shelley. 1982. *Complete Poetical Works*. New York: Modern Library.

Kelly, Sean, and Rosemary Rogers. 1993. *Saints Preserve Us! Everything You Need to Know About Every Saint You'll Ever Need*. New York: Random House.

Kerr, Jessica. 1969. *Shakespeare's Flowers*. New York: Thomas Y. Crowell.

Keyes, Frances Parkinson. 1961. *Rose and the Lily: The Lives and Times of Two South American Saints*. New York: Hawthorn Books.

Keyte, Hugh, and Andrew Parrott, eds. 1992. *Oxford Book of Carols*. Oxford: Oxford University Press.

King-Hele, Desmond, comp. 1994. *Concordance to Erasmus Darwin's Poem "The Botanic Garden."* London: Wellcome Institute for the History of Medicine.

Kirtland, Mrs. C. M., ed. nd. *Poetry of the Flowers*. New York: Thomas Y. Crowell.

Landon, L. E. 1854. *Complete Works of L. E. Landon*. Boston: Phillips, Sampson.

Langland, William. 1978. *Piers Plowman* (C-text). Ed. Derek Pearsall. London: Edward Arnold Ltd.

Language and Poetry of Flowers. nd. New York: Leavitt and Allen Brothers.

Language of Flowers. 1875. New York: Dick and Fitzgerald.

Lehner, Ernest, and Johanna Lehner. 1960. *Folklore and Symbolism of Flowers, Plants and Trees*. New York: Tudor Publishling.

Lehner, Ernest, and Johanna Lehner. 1962. *Folklore and Odysseys of Food and Medicinal Plants*. New York: Farrar, Straus and Giroux.

Leighton, Angela, and Margaret Reynolds, eds. 1995. *Victorian Women Poets*. Oxford: Blackwell.

Liberty Hyde Bailey Hortorium. 1976. *Hortus Third*. New York: Macmillan.

Linnaeus, Carl. 1957 [1753]. *Species Plantarum*. 2 vols. Reprint. Dorking, U.K.: Bartholomew Press.

Li Po. 1965. *Works of Li Po—The Chinese Poet*. Ed. and trans. Shigeyoshi Obata. New York: Paragon Book Reprint.

Longfellow, Henry Wadsworth. 1932. *Poems of Henry Wadworth Longfellow*. Roslyn, New York: Black's Readers Service Company.

Longfellow, Henry Wadsworth. nd. *Poetical Works of Henry Wadsworth Longfellow*. London: Cassel.

Loudon, Mrs. J. W. 1851. *Instructions in Gardening for Ladies*. London: John Murray.

Luick, Karl, ed. 1917. *Sir Degrevant* in *Wiener Beiträge zur Englischen Philologie*, Bd. 47. Wien: W. Braumüller.

Mabberly, D. J. 1997. The Plant-book. Cambridge: Cambridge University Press.

MacCarthy, Denis Florence. 1857. *Bell-Founder and Other Poems*. London: David Bogue.

MacLeod, Malcolm, ed. 1936. *Concordance to the Poems of Robert Herrick*. New York: Oxford University Press.

Martin, Laura C. 1984. *Wildflower Folklore*. Old Saybrook, Connecticut: Globe Pequot Press.

Marvell, Andrew. 1927. *Poems and Letters of Andrew Marvell*, vol. 1. Ed. H. M. Margoliouth. Oxford: Clarendon Press.

McClintock, David. 1966. *Companion to Flowers*. London: G. Bell and Sons.

McClinton, Katherine Morrison. 1944. *Flower Arrangement in the Church*. New York: Morehouse-Gorham.

Medieval Flower Garden. 1994. San Francisco: Chronicle Books.

Milton, John. 1977. *Portable Milton*. Ed. Douglas Bush. New York: Viking Press.

Moir, David Macbeth. 1852. *Poetical Works of David Macbeth Moir*. 2 vols. Ed. Thomas Aird. Edinburgh: William Blackwood and Sons.

Moldenke, Harold N., and Alma N. Moldenke. 1986 [1952]. *Plants of the Bible*. Reprint. New York: Dover.

Moncrieff, Charles Scott, ed. and trans. 1919. *Song of Roland [Chanson de Roland]*. London: Chapman and Hall, Ltd.

Moore, N. Hudson. 1904. *Flower Fables and Fancies*. New York: Frederick G. Hall.

Moore, Thomas. 1860. *Poetical Works of Thomas Moore*. Boston: Crosby, Nichols, Lee.

Morgan, Tom. 1994. *Saints: A Visual Almanac of the Virtuous, Pure, Praiseworthy and Good*. San Francisco: Chronicle Books.

Nendick, Eva. 1971. *Silver Bells and Cockle Shells*. London: Michael Joseph.

Ogden, Scott. 1994. *Garden Bulbs for the South*. Dallas: Taylor Publishing.

Orczy, Baroness. 1964. *Scarlet Pimpernel*. New York: Macmillan.

Ovid. 1970. *Metamorphoses*. Trans. George Sandys, 1621. Lincoln: University of Nebraska Press.

Ovid. 1976. *Metamorphoses*. Trans. John Dryden et al. New York: Garland Publishers.

Oxford English Dictionary (CD-ROM). 1993. Oxford: Oxford University Press.

Parkinson, John, 1991 [1629]. *Garden of Pleasant Flowers (Paradisi in Sole, Paradisus Terrestris)*. Reprint. New York: Dover.

Pater, Walter. 1985. *Marius the Epicurean*. Ed. Michael Levey. New York: Penguin Books.

Patmore, Coventry. 1949. *Poems of Coventry Patmore*. London: Oxford University Press.

Pepys, Samuel. 1970 [1661]. *Diary of Samuel Pepys*. Reprint. Eds. Robert Latham and William Matthews. Los Angeles: University of California Press.

Pickles, Sheila. 1987. *Victorian Posy*. New York: Harmony Books.

Pirie, Mary. nd. *Popular Book on Flowers, Grasses and Shrubs*. London: James Blackwood.

Pliny. 1887. *Natural History of Pliny*. Ed. and trans. John Bostock. London: George Bell and Sons.

Plues, Margaret. 1879. *Rambles in Search of Wild Flowers and How to Distinguish Them*. London: George Bell and Sons.

Poe, Edgar Allan. 1965. *Poems of Edgar Allan Poe*. Ed. Floyd Stovall. Charlottesville: University Press of Virginia.

Poe, Edgar Allan. 1966. *Complete Stories and Poems of Edgar Allan Poe*. Garden City, New York: Doubleday.

Pope, Alexander. 1896. *Poetic Works of Alexander Pope*. London: Macmillan.

Pope, Alexander. 1903. *Complete Poetical Works*. New York: Houghton Mifflin.

Potter, Lois, ed. 1997. *Two Noble Kinsmen* by John Fletcher and William Shakespeare. Walton-on-Thames, Surrey: Thomas Nelson and Sons Ltd.

Powell, Claire. 1979. *Meaning of Flowers, a Garland of Plant Lore and Symbolism from Popular Custom and Literature.* Boulder, Colorado: Shambhala.

Pratt, Anne, and Thomas Miller. nd. *Language of Flowers, the Associations of Flowers, Popular Tales of Flowers.* London: Simpkin, Marshall, Hamilton, Kent.

Project Gutenberg: Library of 194 Electronic Texts (CD-ROM). 1994. Concord, California: Illinois Benedictine College.

Quiller-Couch, Arthur. 1940. *Oxford Book of English Verse, 1250–1918.* New York: Oxford University Press.

Quinn, Vernon. 1939. *Stories and Legends of Garden Flowers.* New York: Frederick A. Stokes.

Rapin, René. 1706. *Of Gardens, A Latin Poem in Four Books* [*De Hortorum Cultura*]. Trans. James Gardiner. London: W. Bowyer.

Rendall, Vernon. 1934. *Wild Flowers in Literature.* London: Scholartis Press.

Richards, Gertrude Moore, ed. 1918. *Melody of Earth: An Anthology of Garden and Nature Poems from Present-Day Poets.* Boston: Houghton Mifflin.

Robbins, Maria Polushkin, ed. 1994. *Gardener's Bouquet of Quotations.* London: Robert Hale.

Rose, Graham, and Peter King. 1986. *Green Words: The Sunday Times Book of Garden Quotations.* London: Quartet.

Rossetti, Christina Georgina. 1904. *Poetical Works of Christina Georgina Rossetti.* London: Macmillan.

Rossetti, Dante Gabriel. 1903. *Poetical Works of Dante Gabriel Rossetti.* London: Ellis and Elvey.

Ruskin, John. 1906. *Works of John Ruskin,* vol. 25. Ed. E. T. Cook. London: George Allen.

Sanders, Jack. 1993. *Hedgemaids and Fairy Candles: The Lives and Lore of North American Wildflowers.* Camden, Maine: Ragged Mountain Press.

Scott, Walter. 1854. *Poetical Works of Sir Walter Scott, Bart.* Edinburgh: Adam and Charles Black.

Scott, Walter. 1923. *Rokeby, The Lord of the Isles, The Bride of Thiermain, and Miscellaneous Poems.* Boston: Houghton Mifflin.

Scott, Walter. 1995. *Rob Roy.* New York: Alfred A. Knopf.

Scott-James, Ann. 1984. *Language of the Garden: A Personal Anthology.* New York: Viking.

Seager, Elizabeth, ed. 1984. *Gardens and Gardeners*. Oxford: Oxford University Press.

Shakespeare, William. 1961. *Complete Works of Shakespeare*. Ed. Hardin Craig. Glenview, Illinois: Scott, Foresman.

Shelley, Percy Bysshe, and John Keats. 1982. *Complete Poetical Works*. New York: Modern Library.

Shosteck, Robert. 1974. *Flowers and Plants: An International Lexicon with Biographical Notes*. New York: Quadrangle.

Skeat, Walter W. ed. nd. *Student's Chaucer*. Oxford: Clarendon Press.

Skinner, Charles M. 1911. *Myths and Legends of Flowers, Trees, Fruits and Plants*. Philadelphia: J. B. Lippincott.

Smith, J. C., and E. D. Selincourt, eds. 1916. *Spenser: Poetical Works*. London: Oxford University Press.

Stearn, William T. 1992. *Stearn's Dictionary of Plant Names for Gardeners*. London: Cassell.

Stevenson, Burton E. 1953. *Home Book of Great Poetry*. New York: Galahad Books.

Stevenson, Robert Louis. 1950. *Collected Poems*. London: Rupert Hart-Davis.

Stevenson, Robert Louis. 1986. *Kidnapped and Catriona*. Oxford: Oxford University Press.

Sullivan, Nancy. 1978. *Treasury of American Poetry*. New York: Barnes and Noble.

Swenson, Allan A. 1994. *Plants of the Bible and How to Grow Them*. New York: Birch Lane Press.

Taylor, Gladys. 1956. *Saints and Their Flowers*. London: A. R. Mowbray.

Taylor, Jennifer, ed. 1991. *Gardener's Quotation Book: A Literary Harvest*. New York: Barnes and Noble.

Tennyson, Alfred, Lord. 1898. *Complete Poetical Works of Tennyson*. Cambridge, Massachusetts: Houghton Mifflin.

Tennyson, Frederick. 1891. *Daphne and Other Poems*. London: Macmillan.

Thackery, William Makepeace. 1940. *Vanity Fair*. New York: Heritage Press.

Thaxter, Celia. 1896. *Poems of Celia Thaxter*. Boston: Houghton Mifflin.

Thaxter, Celia. 1898. *An Island Garden*. Boston: Houghton Mifflin.

Thompson, Francis. 1908. *Selected Poems of Francis Thompson.* New York: Charles Scribner's Sons.

Thoreau, Henry David. 1964. *Collected Poems of Henry David Thoreau.* Baltimore: The Johns Hopkins University Press.

Thoreau, Henry David. 1985. *Weekend on the Concord and Merrimack Rivers; Walden, or Life in the Woods; the Maine Woods; Cape Cod.* New York: Library of America.

Todd, Pamela. 1993. *Forget-Me-Not: A Floral Treasury.* Boston: Little, Brown.

Turner, W. J. 1920. *Dark Wind.* New York: E. P. Dutton.

Turner, William. 1989 [1562]. *New Herball,* Part 1. Reprint. Ed. George T. L. Chapman. Cambridge: University of Cambridge Press Syndicate.

Turner, William. 1995 [1562]. *New Herball,* Part 2. Reprint. Ed. George T. L. Chapman. Cambridge: University of Cambridge Press Syndicate.

Tynan, Katherine, and Frances Maitland. 1909. *Book of Flowers.* London: Smith, Elder.

Tyssen, Alicia M. 1894. Fifteenth Century Treatise on Gardening by John Gardiner. Archaelogia 54:157–172.

Vickery, Roy. 1997. *Dictionary of Plant Lore.* Oxford: Oxford University Press.

Virgil. 1903. *Works of Virgil.* Trans. John Dryden. London: Oxford University Press.

Virgil. 1915. *Georgics and Eclogues of Virgil.* Trans. T. C. Williams. Cambridge, Massachusetts: Harvard University Press.

Walafrid Strabo. 1966. *Hortulus.* Trans. Raef Payne. Pittsburgh: Hunt Botanical Library.

Waley, Arthur, ed. 1919. *A Hundred and Seventy Chinese Poems.* New York: Alfred A. Knopf.

Walton, Izaak. 1996. *Compleat Angler.* New York: Modern Library.

Ward, Mrs. Humphry. 1895. *Story of Bessie Costrell.* New York: Macmillan.

Wavell, A. P. 1945. *Other Men's Flowers: An Anthology of Poetry.* New York: G. P. Putnam's Sons.

Wharton, Edith. 1968. *Collected Short Stories of Edith Wharton,* vol. 1. Ed. R. W. B. Lewis. New York: Charles Scribner's Sons.

White, Gilbert. 1875. *Natural History and Antiquities of Selbourne.* London: Macmillan.

White, Gilbert. 1982. *Journals of Gilbert White.* London: Futura Publications.

Wilde, Oscar. 1915. *Novels and Fairy Tales of Oscar Wilde.* Ed. H. S. Nichols. New York: J. J. Little and Ives.

Wilde, Oscar. 1940. *Best Known Works of Oscar Wilde.* New York: Book League of America.

Wilde, Oscar. 1982. *The Picture of Dorian Gray and Other Writings.* Ed. Richard Ellmann. New York: Bantam Books.

Wilder, Louise Beebe. 1936. *Adventures with Hardy Bulbs.* New York: Collier Books.

Wither, George. 1871. *Juvenilia: Poems of George Wither.* Manchester: Spenser Society.

Wordsworth, William. 1950. *Selected Poetry.* New York: Modern Library.

Wordsworth, William. 1977. *Poems.* 2 vols. Ed. John O. Hayden. New York: Penguin Books.

Wordsworth, William. 1982. *Poetical Works of William Wordsworth.* Ed. Paul D. Sheats. Boston: Houghton Mifflin.

Zheleznova, Irina, ed. 1983. *Russian 19th Century Verse: Selected Poems.* Moscow: Raduga Publishers.

INDEX OF PERSONS

Titles of works quoted in this volume are listed under the corresponding author's name. Parenthetical references following those titles refer to the sources as they are listed in the Bibliography. Any titles lacking a parenthetical reference are listed in the Bibliography according to the author's name.

Swithin (saint) (d. 862 CE), 397–398

Sylvester, Josuah (1563–1618)
Du Bartas' Triumph of Faith, the Sacrifice of Issac (Oxford English Dictionary 1993), 282

Tannahill, Robert (1774–1810)
"The Midges Dance Aboon the Burn" (Bryant 1970), 204–205

Tennyson, Alfred, Lord (1809–1892)
"Becket" (Oxford English Dictionary 1993), 184
"The Brook" (Oxford English Dictionary 1993), 145, 267
"The Daisy," 43
"Flower in the Crannied Wall" (Harvey 1991), 370
The Idylls of the King, "The Last Tournament," 337; "Marriage of Geraint," 257
"In Memoriam," 151–152, 194–195, 340
"Lady Clare" (Ferris 1957), 247–248
"Last Town," 132
"The Lotos-Eaters" (Great Literature 1992), 29, 373
"Maud," 19, 178
"Song: A Spirit Haunts," 199, 348

Tennyson, Frederick (1807–1898)
"Psyche," 361

Teresa of Lisieux (saint) (1873–1897), 399–400

Thackeray, William Makepeace (1811–1863)
Vanity Fair, 351

Thaxter, Celia (1835–1894)
"August" (Thaxter 1896), 375
An Island Garden (Thaxter 1898), 201, 336
"My Hollyhock" (Thaxter 1896), 201

Theocritus (ca. 316–260 BCE)
"The Cup" (Higham and Bowra 1938), 223

Thomas, Edith Matilda (b. 1854)
"Frost To-night" (Richards 1918), 120

Thompson, Francis (1859–1907)
"The Poppy," 299

Tomson, Graham R. See Watson, Rosamund Marriott

Thomson, James (1700–1748)
The Seasons (Hunt 1993), 253, 336, 356

Thoreau, Henry David (1817–1862)
"Buttercups and geraniums cover the meadow" (Durant 1976), 166
Cape Cod (Thoreau 1985), 132
"I Am a Parcel of Vain Strivings Tied" ("Sic Vita") (Thoreau 1964), 377

Tickell, Thomas (1685–1740)
"Kensington Garden" (Rendall 1934), 335–336

Timrod, Henry (1828–1867)
"Ode" (Sullivan 1978), 235–236

Tolstoy, Alexei (1817–1875)
"Crimean Sketches" (Zheleznova 1983), 226

Tradescant, John (1608–1662), 53–54

Turner, William (1508–1568)
The Names of Herbes (Oxford English Dictionary 1993), 114, 131, 343
New Herball, 109, 283
"Spurge purgeth thynne fleme vehemently" (Oxford English Dictionary 1993), 131

Twamley, L. A. (1812–1895)
"Why, when so many fairer shine" (Flower Lore nd.), 95

Urban of Langres (saint) (d. 230 CE), 400

INDEX OF PLANTS

COLOPHON

This book is set in Centaur, a typeface developed in 1929 by Bruce Rogers and Frederic Warde. It is based on the Renaissance roman type of Nicholas Jenson and the Renaissance italic font of calligrapher Ludovico degli Arrighi and was digitized by Monotype Typography Inc. This book was designed, typeset, and laid out by Susan Applegate on a Power MacIntosh 8400 in QuarkXpress, quite a different way of putting together a book from the Renaissance letterpress printing of Jenson and Arrighi.